A GUIDE

TO

PRACTICAL CHEMISTRY

By

Adams Guscan

Cover design by Caleb Okello

Ideal Impressions Ltd (U)

A Guide to Practical Chemistry

Adams Guscan

+256703784952

adams.guscan@gmail.com

DEDICATION

To all students yearning to become better citizens through application of scientific concepts, practice will enhance your learning.

ACKNOWLEDGEMENT

Great thanks to the almighty God who has watched over me all the days of my life. I also thank all those who taught me in school for the impeccable job they did. Special thanks to my science teachers Mr. Steven of Karambi Senior secondary school, Mr. Mutuya of Mengo Senior School and Dr. G.W. Bazirake of Kyambogo University.

Great indebtedness is also attributed to Mr. Filly Lawrence for his major contribution in typing and editing this work.

I thank the management of St. Paul S.S and especially the head of Chemistry department for allowing me use their laboratories for most of the experiments looked at here.

INTRODUCTION

Welcome to the *Guide to Practical Chemistry*! This book is primarily intended for any student preparing to sit for 'O' level examinations. It is equally useful to students offering Chemistry subject at 'A' level.

Chemistry is just one of the subjects you have to study, and therefore you have limited time for each subject. Keeping in mind your limited time and your tight timetable, I have tried by all means to see that this work fulfills your requirements and fits in your time frame.

The biggest mistake most students do is, to develop a bad attitude towards the subject or the subject teacher. We are all capable of succeeding, but the way we set up our minds towards what we are meant to do, determines our destiny. With a good attitude, and a clear mind, all subjects are passable once given time.

I believe you are going to benefit a lot from this book if you thoroughly read it. The special features of this book are;

1. Laboratory techniques: I have basically looked at safety precautions which you must take while in the laboratory
2. General laboratory directions: I have looked at the different apparatus used and sources of errors during practical observations and reporting of results.
3. Volumetric analysis: I have looked at recording, presentation and treatment of results, indicators, and some examples on volumetric analysis, basically looked at Acid-Base titrations.
4. Qualitative Analysis: I have only written about inorganic qualitative analysis and some examples.

I would like to suggest to you three important points so that you can take maximum advantage of this book.

- Read and practice qualitative analysis very well before your examination time, make it part of your language, do not try to cram color changes, do not cram reactions but understand reactions. Make this your rule, understand what brings about color change and which elements are associated with which color changes. For example, understand why some substances dissolve in water where as other do not.

- Do not just cram the reagents used or examples and answers given in this book, but derive maximum advantage by following the steps given here in when a related number appears in your exam.

- Read and understand this work thoroughly and consistently. It can be your wonderful step on your journey to success, and the world of science is waiting for you.

Concentration begins with commitment; the more we are committed to something, the easier it is to concentrate on it. The postage stamp is useful because it has the ability to stick on the document or anything on which it is put until it gets there. A person of excellence is one who is willing to stick to his/her goals until he sees the outcome. Without commitment, we float aimlessly from activity to another from one subject to another, from one subject combination to another or from one course to another.

Your job as a student now is to read books. Read your books and use this wonderful time profitably. You have a lot ahead of you, and all of them need a strong foundation, which you are meant to build now. May you find this book very helpful to you on your journey to a world of science. Once again welcome to the *Guide to Practical Chemistry*!

Adams Guscan

Table of Contents

PART ONE

LABORATORY TECHNIQUES

1.1 Safety

A number of accidents do occur even in well-run and organized laboratories. It's therefore essential to ensure that prior preparations should be made to deal with any emergency cases. In this book, I have discussed only major parts that need to be soundly learnt by every student. I however believe, chemistry teachers and laboratory technicians will enlarge this topic.

Safety in every laboratory is an area that has to be given due consideration. It's your responsibility to ensure that you are safe and also those around you in the laboratory. As long as you know the dangers associated with the chemical, and other apparatus you are handling while in the laboratory, safety becomes a common sense.

Precautions to Take;

It's not good to think of safety when something starts to go wrong. It's highly advisable to be safety conscious all the time; this will help you to know what actions to take if an accident occurs. For example you must know the location of the water tap, incase your body accidentally comes in contact with an acid. You must know the position of the fire extinguisher incase fires sets.

A well-lit, and ventilated laboratory with proper arrangement of worktables minimizes on occurrence of accidents. That is however not enough; you must know the hazards associated with each experimental phase and take proper precautions. You may need to be very careful while in a chemistry laboratory than in a physics laboratory. The dangers associated with combustion of organic solvents, combustible gases and poisoning with hazardous chemicals may be severe than those associated with broken glasses.

1.1.1 Broken glasses

Although a number of students ignore this, broken glasses are common cause of accidents in especially chemistry laboratories. Learning the correct ways of handling glass apparatus is the only way to avoid accidents associated with breaking glasses. Inspection of apparatus for cracks or deep scratches before use is a necessary measure that can avoid accidents caused by broken glass. When fitting glass tubing and thermometers to corks always protect your hands with a cloth, or hard leather made hand gloves.

Glass is produced and used in vast quantities and there is considerable variation in quality. Large glass containers must not be handled by the neck and those keeping reagents need special care in this respect. Reagents in glass bottles must not be stored in direct sunlight or warm floors. It should be remembered that, broken glasses need to be removed as soon as possible by putting them in a clearly marked container used exclusively for that purpose. Cracked glassware should not be used in laboratory experiments

Pipettes and burettes can be dangerous if they are not properly used. Mouth operated pipettes are inexpensive to purchase but are dangerous especially if the user is new to them. A student has to be very careful not to suck the contents into his/her mouth, as some if not most of the reagents used in chemical laboratories are very dangerous to the body. If a student fails to fully immerse the sucking point of a pipette into the sample reagent, s/he is most likely to suck it into the mouth. Pipette fillers are readily available and they overcome the hazards that arise with use of a mouth-operated pipette.

Mouth operated pipettes must never be used for drawing volatile liquids, aqueous ammonia, concentrated acids or alkalis, and toxic liquids. Pipettes should be washed immediately after use and then stored in racks.

When properly clamped, the top of a burette is likely to be above the head of the operator. The burette should be released and brought down below eye level before any attempt to fill it using a funnel. Students should be forbidden to climb on stools to fill burettes, as this is another common cause of accidents in a chemistry laboratory.

All glassware should be inspected regularly for flaws and any that are detected should be repaired immediately or the item rejected. On no account should laboratory glassware be used as drinking vessels, remember that, glass is not totally an inert material, it should therefore be cleaned as soon as possible every after use.

1.1.2 Use of Gas Cylinders

Gas cylinders are widely used in schools and most accidents involving cylinders arise from inadequate storage or misuse. Gas cylinders must always be clamped in right positions or laid horizontally on the floor and properly prevented from rolling. They must not be subjected to heat from either sunlight or any external heat source. They should never be exposed to corrosive fumes.

A student should learn how to use a gas cylinder with the help of either his/her teacher or a laboratory technician. Before use, the valve of every cylinder should be checked. Cylinder valves must always be kept smooth, and opened slowly. A harmer should never be used to exert pressure on the standard key of a gas cylinder. If hand pressure is insufficient to open a main cylinder valve, the cylinder should be returned to the supplier with an explanatory note. No attempt must ever be made to oil or grease a cylinder valve and valves suspected of leakage should be tested with soap water.

Gas cylinders contain different contents, there are those with combustible gases like methane, which are commonly used to produce heat in the laboratory, and there are those which carry non-combustible gases like carbon dioxide, which actually do not support combustion.

In all cases, the main cylinder keys should be fastened to the cylinder or trolley so that it may be used to cut off the supply of gas from the cylinder in an emergency. A gas cylinder must never be connected to any apparatus without first establishing the contents of the cylinder, establish whether it contains a combustible or non combustible substance and then control the rate of gas flow. If you cant do this, then you highly risk burning your laboratory.

Gas cylinders must not be stored in a closed area that would allow a high concentration of leaking gas to develop.

The main hazards associated with the use of gas cylinders containing large volumes of gases stem from the high pressure of the gas and from the chemical properties of the contents, many oxygen cylinders for example are filled at a pressure as high as 200 atm. A reducing valve and gauge must therefore be fitted to the cylinder to bring the flow rate and pressure of the escaping gas down to a safe level

Compressed gas cylinders must always be treated with care. Valves should be opened and closed only with cylinder keys provided. It is extremely dangerous to direct a set of compressed air onto the skin especially where there are wounds or cuts. This can result in painful injuries or even death. After knowing how to handle a gas cylinder, the next question that may come to your mind can be; how do I differentiate gas cylinders, for example, "how can I know that this cylinder uses Nitrogen, or carbon dioxide?"

Identification of Cylinder Contents

Gas cylinders are color coded to indicate their contents as shown below.

Gas	Color	Characteristics
Oxygen	Black	Odorless, doesn't burn but supports combustion, no oil, grease or lubricant should come in contact with gas cylinder
Nitrogen	Grey	
Hydrogen	Red	Odorless, burns with an invisible flame, collects fast in any enclosed area.
Chlorine	Yellow	
Carbon dioxide	Grey/green	Can cause nose to sting, does not support combustion and is thus a common fire extinguisher.
Air	Grey	

Table 1. Showing colors associated with cylinder contents

Hazards Associated with Contents of the Compressed Gas Cylinders

Gas	Hazard	Notes
Oxygen	Assists fires, and can cause explosion, it's an oxidizing	Oxygen can strongly assist a fire once it has started. Compressed Oxygen must not be allowed to come into contact with grease, oil or other combustible materials or violent explosions may result.
Chlorine	Poisoning	Cylinders containing poisonous gases like Chlorine and carbon monoxide must not be stored in the laboratory incases of leaks
Hydrogen	Fires and explosions	Hydrogen is highly flammable gas and forms explosive mixture with oxygen or air.
Carbon dioxide		

Table 2. Showing hazards of cylinder contents

1.1.3 Fire Precautions

Waste papers should not be used for lightning Bunsen burners. It is recommended to use splints and the splint must be properly extinguished before being discarded. When using low boiling solvents like ethanol, ether, propane or petroleum spirit, extra care must be taken when handling these solvents. No burners should be nearby. You must always know where the fire extinguishers are kept and how to operate them when fires do occur.

Fire can put down a laboratory within a short period of time because some chemicals and gases used in a chemistry laboratory accelerate combustion. Bunsen burners present serious fire hazards because they produce naked flames that burn at higher temperatures, and as a result, there is potential for an accident to occur during use of a Bunsen burner. For your safety and for the safety of others, it's very important that you take maximum precautions, and incase of any fire, put on the fire alarm, alert the laboratory attendant and evacuate the laboratory immediately. Ensure the laboratory is well ventilated, with maximum light. Every laboratory should have emergency exists.

1.1.4 Hazardous Chemicals

In most cases, serious dangers of hazardous chemicals only arise when the chemical is repeatedly exposed to the skin (body). The table below shows the list of the most common chemicals known to be hazardous.

Hazardous Chemical	Comments:
Benzene	Toxic vapor causing dizziness
Ether	Very readily inflammable
Ethane Diamine	Irritant and harmful by skin absorption
Hydrazine	Corrosive
Nitro benzene	Toxic by vapor and skin absorption
Amiline and aromatic amines	Toxic by vapor and skin absorption, may be carcinogenic
Chlorinated alkanes	Most of these are narcotic causing mental confusion
SO_2, H_2, SO_4, NO_2, CL_2 Br_2, I_2, HF	All these cause rapid destruction of the skin when concentrated. HF is most dangerous

Table 3. Showing hazardous Chemicals

Note: It is always important to check bottle labels before using a chemical to avoid dangers with chemicals. Use a safety pipette to measure samples if solution is volatile or concentrated. Do not suck volatile solutions with your mouth.

Do not eat, drink or taste chemicals when in the laboratory. Fume cupboards must always be used if experiments being carried out involve hazardous volatile chemicals. Fume cupboards remove all poisonous gases.

Bottles of concentrated Ammonia, anhydrous Aluminium Chloride should be kept cool and never brought near radiators or under direct sunlight. When accidents occur on opening containers of volatile chemicals, always use first aid emergency treatment. There is no substitute for professional medical examination at the earliest possible time. Seek professional medical treatment as early as possible. Do not only rely on first aid treatment incase of an accident.

If a bottle containing any compound lacks any means of identification of its contents, then either the technician, or the teacher responsible MUST dispose off the contents of that bottle.

1.2 First Aid Emergency Treatment

1.2.1. Burns.

All burns caused by heating should be treated by immersion of the effected part in cold water and then apply a cold wet dressing. However, the treatment of a chemical burn varies with the chemical that caused it, though washing with plenty of water is the first solution. The table below shows special washes for use that should be available in the first aid cupboard.

Chemical causing burn	Neutralizing Wash
Acids eg HF	Use 2M Ammonium carbonate for the case of HF. Use this wash and take the victim to the hospital for more treatment. For other acids including HCL use plenty of water to wash off the acid
Alkalis	Use Ethanoic aid
Phenol	Use ethanol and take the victim to the hospital for treatment
Bromine	Use 2M ammonia

Table 4. Showing Chemicals that cause burns, and their neutralizing washes

Note: When an accident occurs, even if its minor, a crowd of onlookers is undesirable. Give the victim a free atmosphere.

1.2.2 Swallowed Poison.

If poison is swallowed, give the victim plenty of water at once if he /she is conscious. If the poison swallowed is corrosive, give limewater (calcium hydroxide) as soon as possible. When an accident results into a respiratory problem as in cases of gassing, knowledge of artificial respiration methods is required.

1.2.3. Cuts.

When one gets an injury/cut, however minor it could be, she/he should be seated down for treatment as some victims faint even with minor cuts. Wash the wound properly with cold water to remove any foreign bodies, apply an antiseptic cream and protect the effected part with a suitable dressing.

Some Important Points to Remember

- Always check the name on a bottle of the chemical you exactly want to use.
- Never smell gasses directly. Be very cautious when drawing fumes of the gas towards your nose.
- Never point a test tube containing chemicals, which you are heating towards yourself or any one else and never look directly down into the tube. Do not bring your nose directly closer to the contents being heated.
- Always follow procedure one after the other, and never perform un authorized experiments.
- Never eat or drink while in the laboratory
- Never taste anything unless instructed to do so
- Always work steadily and without haste
- Never get your clothes, paper or hand near a Bunsen burner flame.
- Where possible always put on a laboratory coat while performing any experiment.
- Never try to force glass tubing when putting it into or removing it from the cork.
- Never remove chemicals from the laboratory
- Always wash your hands after a practical experiment
- Always handle flammable liquids like ethanol, propane, with great care, and keep them from naked flames.
- Where possible, wear eye, nose or face protections when handling acids or alkalis or when heating chemicals and performing exothermic reactions.

1.3. Filtration

There are many techniques used for the separation of a liquid or solution from a solid. However I will only look at two techniques;

Simple filtration: Here we use a filter funnel and a piece of filter paper folded into four. It's usually reserved for inorganic substances. Candidates always make mistakes during filtration because of lack of prior knowledge on how to fold a filter paper into the funnel. A candidate must ensure that the paper is really carefully folded and fitted carefully into the funnel and wetted thoroughly well with water, or the appropriate solvent before filtration is started.

The contents of the filter paper should not reach within half an inch of the top of the paper. Solid impurities can be removed from a liquid (mainly organic liquids) by using a fluted filter paper. This paper allows the whole of its parts to be active rather than half of it. The easiest ways of folding a filter paper are illustrated below.

Diagrammatic Illustrations of Folding a Filter Paper

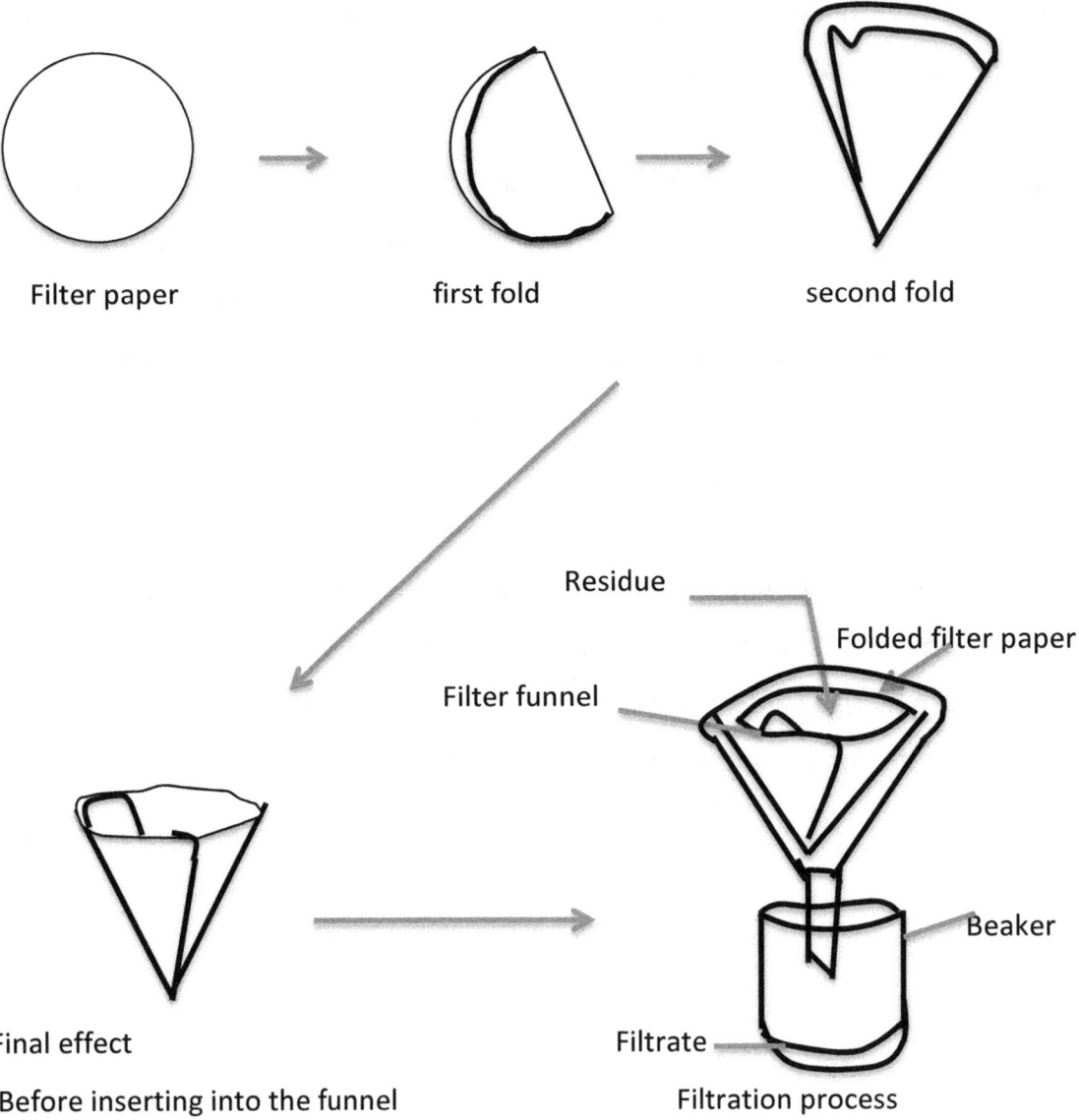

Filter paper   first fold   second fold

Residue

Folded filter paper

Filter funnel

Beaker

Final effect

Before inserting into the funnel

Filtrate

Filtration process

Distilled water can be poured in before use. A clean finger applied cautiously to prevent tearing the wet paper is used to smooth the paper and obtain a tight seal of the paper to glass at the top. In this way, filtration can be facilitated.

1.3.1 Washing the Residue

In most cases, one may be instructed to wash the residue with water or with wash solution. The simplest technique is pouring water on top of the residue for about three times and allowing the filtrate to pass through the filter or the residue can be stirred with water in the beaker, or any wash solution as instructed and the washings decanted through the filter. This procedure is repeated for about three times.

There are some other techniques of filtration such as gravimetric filtration, and Bucher funnel filter pump. However, these have not been looked at here in this book.

1.4. General Cleaning

This is one of the most important aspects of practical work, which is often neglected. Glassware become very difficult to clean once kept before cleaning, or if allowed to get dirty on storage. This may require expensive chemicals to clean where as water would be sufficient to clean them if used at the right time. Cleaning of apparatus should be done immediately after use. All glassware should be rinsed with water immediately after using them. This will remove all substances that would otherwise stick on glass and make it hard to clean it.

Normally a communal washing place is provided with assortment of beakers, a few filter funnels, a bottle of suitable washing solvents, and a residue bottle into which all washings are poured. Commonly ethanol and propane are used, however, only one solvent should be used at a time during communal washing.

For flasks, which get stained with difficult materials, concentrated Nitric Acid is used to remove them. Your teacher or technician can lead you through a special procedure for using concentrated Nitric Acid for washing flasks.

An acid bath can be used to clean small pieces of glassware, which are really difficult to clean. This mixture consists of concentrated Sulphuric acid with a little nitric acid.

PART TWO

GENERAL LABORATORY GUIDELINES

As already discussed in part one, a student must develop independent judgment in connection with his/her lab work. Common sense plus awareness of the danger spots are the main requirements of the beginning student in this regard.

2. 1. Neatness and Cleanliness

A good analyst is always neat. A student with an orderly work-table, is not likely to mix up samples or add wrong reagents. S/he will not break glasses or spill solutions.

Neatness in the laboratory must extend from the students' own worktable, or bench to the shelves where laboratory materials are kept. Glassware that look clean may or may not be clean. Therefore do not presume something is clean by just looking at it. Clean it yourself, that's when you will be sure it's clean. Surfaces on which no visible dirt appears are often still contaminated by thin invisible greasy material so ensure you thoroughly clean your worktables.

2.2. Apparatus

There are simple equipment found in the Chemistry laboratory, which may seem useless to a layman but very important to the analyst. Where as I will not look at one by one, I will look only at the most common ones and these include;

Pipettes: These are of various forms e.g, transfer pipettes, measuring pipettes, and lambda pipettes. The transfer pipette is used to transfer an accurately known volume of solution from one container to another. They normally come in two different volumes of 20m, and 25ml. Before using a pipette, carefully note which volume it is, is it a 25ml, or a 20ml? The pipette should be cleaned first before being used

Transfer pipette 20/25ml at 25°c

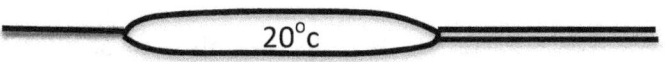

Filling pipette, liquid drawn above graduation mark, and use of fore finger to adjust liquid level in pipette before transferring into the conical flask

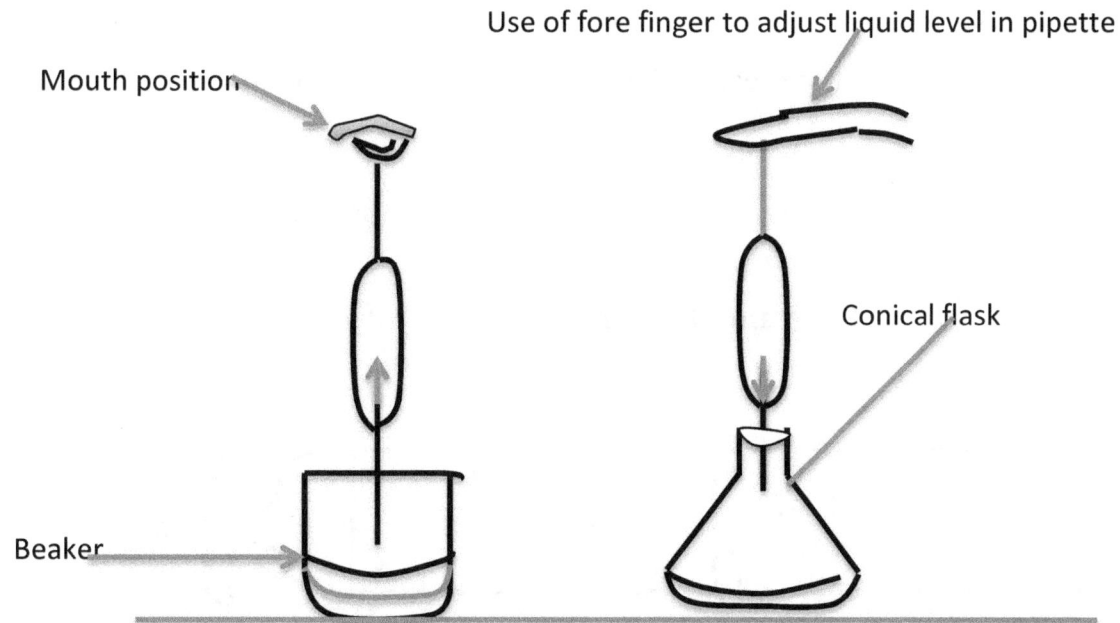

Note: After drawing the solution from its container into the pipette, the contents of the pipette are then allowed to run into the desired container with care being taken to avoid splattering. With the pipette in the vertical position, the solution is allowed to drain into the desired container. A small volume of solution will remain in the tip of the pipette but the pipette has been calibrated to take this into account. Therefore this small quantity of the solution should not be blown out.

Burettes: The burette is used to deliver accurately known but variable volumes in titrations.   When using a burette, ensure that it is clean and always be very cautious in reading burette values.   Much practice is needed in order to become familiar with the graduations and the estimations between them.

An ordinary 50ml burette is graduated in 0.1ml intervals and should be read to the correct nearest 100th of ml.   Do not leave solutions standing in the burette for long periods of time.   After each experiment, the contents in the burette should be discarded and the burette properly cleaned with distilled water and then stored (left to dry up).

Note: Aqueous solutions in a burette forms a concave surface referred to as a meniscus, when the solution is not colored, the bottom of the meniscus is ordinarily read.

Diagram of a burette

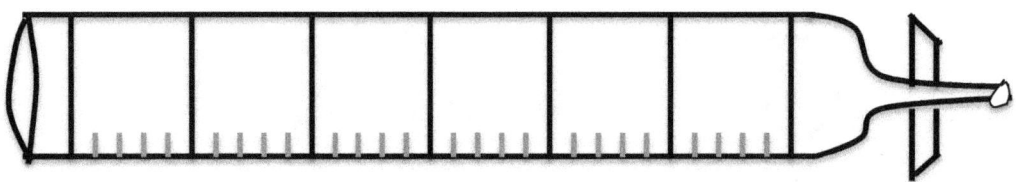

In the picture below, the reading is 1.10 cm^3 the eye must be positioned at the meniscus of the liquid. Check whether the liquid has a lower or upper meniscus

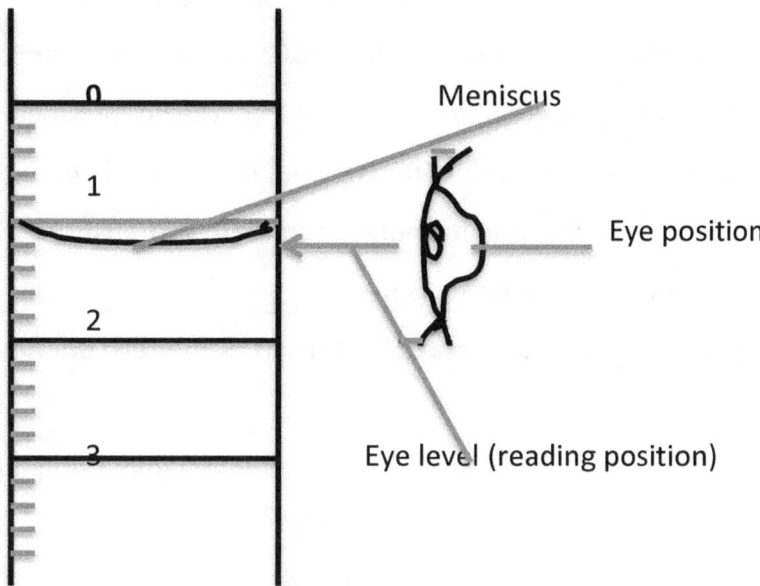

Ensure that there are no air bubbles in the tip of the burette before a titration is started otherwise, such air bubbles would cause errors.

Bunsen burner

The Bunsen burner is commonly used in school laboratories to produce heat when carrying out experiments that require heating. It uses a natural gas which is basically methane gas as it's fuel. It can also use other gases such as propane, butane or a mixture of both. Methane burns in either plenty of oxygen, or with limited oxygen. The Bunsen burner has holes, which can be opened or closed producing two different flames. A very hot blue flame (non-luminous flame) is produced when the air holes of a Bunsen burner are opened. This means there is complete combustion.

Methane + Oxygen $\rightarrow$ Carbon dioxide + water.

A blue flame is produced when air holes of a Bunsen burner are opened

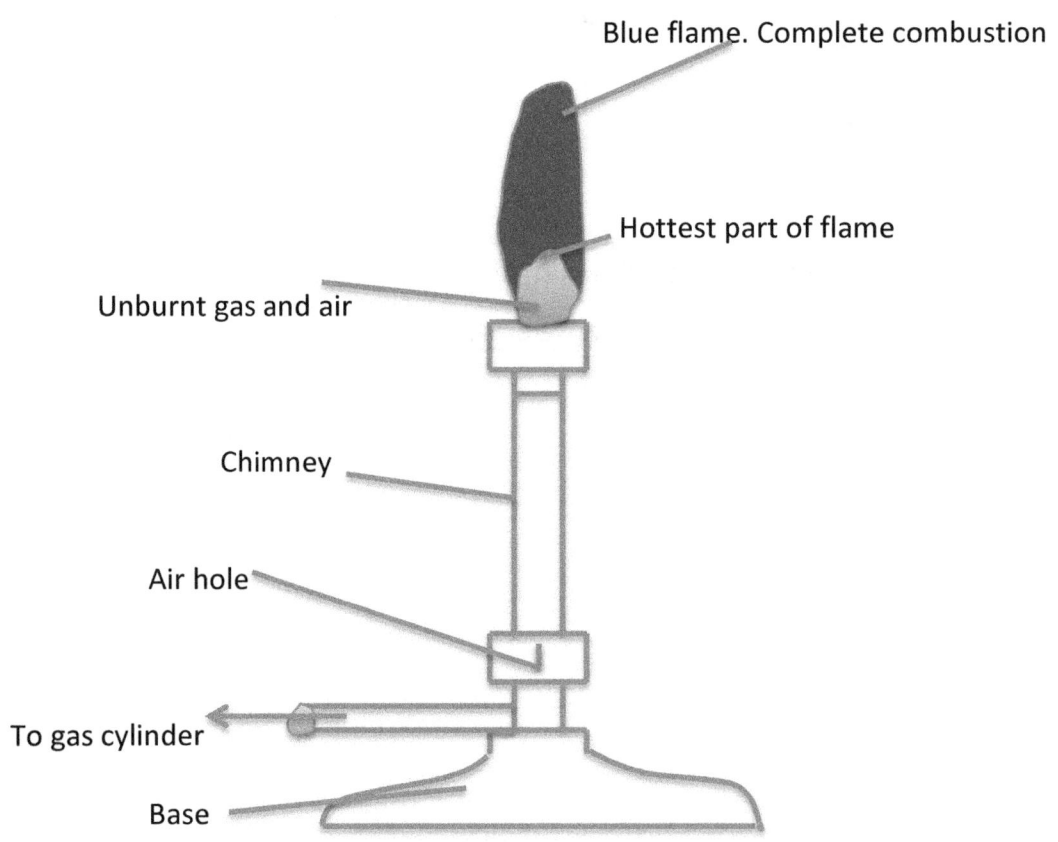

Diagram of a Bunsen burner with air hole open

The hottest part of the flame is at the tip of the dark blue flame

When the air holes of a Bunsen burner are closed, methane gas can only mix with air at the tip of the gas barrel. As a result, an incomplete combustion occurs producing different products compared to when the gas burns in plenty of air.

Methane + Oxygen $\rightarrow$ Carbon monoxide + Carbon + water

A yellow flame (luminous flame) is produced. This flame produces less heat energy than the blue (non-luminous flame).

A yellow flame is produced when air hole of Bunsen burner is closed

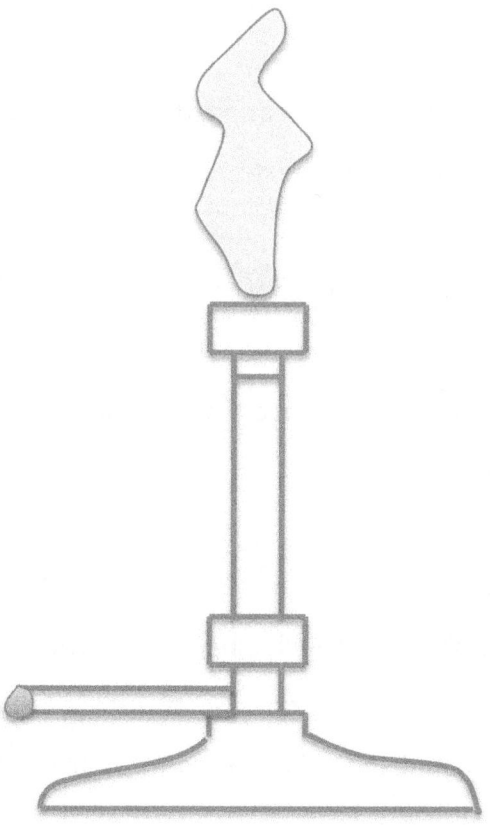

Diagram of a Bunsen burner with air hole closed

Can you draw the differences between a luminous and a nonluminous flame?

It is normally recommended to use a safety flame if you are not familiar with use of a Bunsen burner.

Precautions when using a Bunsen burner

- Do not transfer the Bunsen burner when it's in use.
- Remove all paper, books, cloth or any combustible material from the worktable
- Tie any long hair, cloth, or anything on your hands that can cause accidents during heating.
- Do not light the Bunsen burner under wooden shelves, plastic boxes, or near paper bags. Put it away at least 12 inches from the access area.
- Ensure the gas hose fits securely on the gas valve before lighting the Bunsen burner
- If a glass is being used during heating, do not touch it with naked hands.
- Do not shut off gas before heating is done, and immediately shut it off after experiment.
- Let other people in the laboratory get to know that the Bunsen burner is in use.
- Never leave the laboratory while the Bunsen burner is on.
- Ensure that the burner is placed on a leveled surface to avoid pouring and causing accidents.

There are other equipment used including, volumetric flasks, desiccators, test tubes, stirring rods, wash bottles, tripod stands, measuring cylinders, and many others. A student is highly encouraged to be familiar with all these apparatus before entering a chemistry practical exam.

PART THREE

VOLUMETRIC ANALYSIS

This is a technique of analysis in which the substance to be determined in the solution form is made to react with a measured volume of solution of known concentration. It is therefore necessary to establish exactly the end point of the reaction. For suitable applications of volumetric methods a process must satisfy the following requirements.

- The reaction must be rapid so that little time is lost performing the titration.
- There must be a suitable indicator to mark the end point of the reaction
- A suitable standard solution must be available as a titrant for carrying out the reaction
- The reaction must be complete when equivalent amounts of reactants have been brought together from where we can make calculations basing on the volumetric data.
- There should be only one single reaction occurring between the desired constituent and the titrant (Ensure that the solutions used do not contain any kind of impurities)

3.1 Key Terms Used in Volumetric Analysis

(a) **Standard solution**: This is a solution whose concentration has been accurately determined.

(b) **Primary Solution:** This is a substance available in a pure form or in a state of known purity, which is used to standardize other solutions.

(c) **Molarity**: This refers to the number of moles of solute per liter of solution

(d) **Mole:** This is the number of grams of a substance containing as many elementary particles (entities) as there are atoms in 12g of carbon -12 isotope

(e) **End point**: This is the point in a titration where there is a sudden change in a physical property of the solution.

(f) **Indicator:** This is a chemical substance, which exhibits various colors in the presence of excess analyte or titrant.

(g) **Titrant**: This is the reagent (standard solution) which is added from a burette to react with the analyst

(h) **Titration**: This is the process of measuring the volume of titrant required to reach the equivalence point.

(i) **Equivalence point:** The point in a titration where chemically equivalent amounts of analyte and titrant are present.

(j) **Formula weight**: This is the summation of atomic weights of all the atoms in the chemical formula of a substance.

3.2 Determination of Equivalence Point (End Point)

Neutralization is not accomplished by visual changes e.g alteration of the color of the solution, therefore a suitable indicator MUST be added to the titrated solution (one in the flask) in order to determine the end point.

The end point in neutralization reaction occurs at a given P.H valve. All indicators show different colors in accordance with P.H (color of indicator depends on P.H) and such indicators can be described as P.H indicators.

The most commonly used indicator used at 'O' level are phenolphthalein (colorless) and Methyl orange (yellow) indicators.

The table below shows a list of some Acid base indicators with their expected color changes and the two P.H media.

Indicator	Color Change		P.H range
	Acid	Alkali	
Phenolphthalein	Colorless	Red	8.0-9.6
Methyl Orange	Red	Yellow	2.5-4.4
Methyl red	Red	Yellow	4.2-6.2
Methyl purple	Purple	Green	4.8-5.4
Methyl yellow	Red	Yellow	2.9-4.0
Litmus	Red	Blue	4.5-8.3
BromoPhenol blue	Yellow	Blue	3.0- 4.6
Bromocrestol purple	Yellow	Purple	5.2 – 6.8
Bromolthymol blue	Yellow	Blue	6.0-7.6
Phenol red	Yellow	Red	6.8 – 8.4
Thymol blue	Red	Yellow	1.2 – 2.8
Alizarin yellow	Yellow	Violet	10.1 – 12.0

Table 5: Showing Acid-base indicators, their expected color changes, and P.H ranges

The choice of an indicator of an indicator to mark the end point in a titration is governed by the P.H of the reacting solutions. Generally one should select an indicator, which changes color at approximately the P.H at equivalence point of the titration.

For weak acids, the P.H at the equivalence point is above 7 and phenolphthalein is the best choice. For weak bases, the P.H is below 7 at the equivalence point; Methyl red or Methyl orange is widely used.

For strong acids and strong bases, methyl red, and phenolphthalein are suitable choices.

3.3 Recording Experimental Results

All data (results) obtained in the laboratory during an experiment should be recorded directly in the table of results at the time they are obtained. Most forbidden is the recording of data on loose papers with the idea of copying it into the table of results later. Remember, you have limited time in the examination room.

All values should be recorded in ink. If a mistake is made and a recorded value is invalidated, it is not to be rubbed but it is crossed out once.

Record your values in away that can help you to easily locate errors in calculations and to orderly present your data. Make the use of table of results always given to you as shown below.

Number of Titrations	Rough	1	2	3
Final burette reading (cm^3)				
Initial burette reading (cm^3)				
Volume of titrant (cm^3)				

Table 6. Showing data entry during volumetric analysis

3.4. Examples of Experiments

3.4.1 Experiment 1.

TO DETERMINE THE NUMBER OF MOLECULES OF WATER OF CRYSTALISATION IN WASHING SODA ($Na_2CO_3 \, nH_2O$)

Provided:

- BA_1 = 0.1M HCl
- BA_2
- Methyl orange indicator
- $25cm^3$ pipette,
- $50cm^3$ burette
- A stand and clamp
- 3 conical flasks

End point: Yellow to red (Acid solution in the burette)

Reaction

$Na_2CO_3 \, nH_2O_{(aq)} + 2HCl_{(aq)} \rightarrow 2NaCl_{(aq)} + (n+1) \, H_2O_{(l)} + CO_{2 \, (g)}$

Ionic equation

$CO_3^{2-}{}_{(aq)} + 2H^+{}_{(aq)} \rightarrow H_2O_{(l)} + CO_{2 \, (g)}$

Procedure

Rinse and fill the burette with solution BA_1. Pipette out $25cm^3$ of BA_2 into a conical flask. Add 1-2 drops of Methyl orange indicator. Add BA_1 solution from the burette until a permanent red color is obtained. Repeat the above procedure until you get three consistent results.

Note:

BA_2 is a solution made by dissolving 11.4g of $Na_2CO_3 nH_2O$ in 1 liter of water

Recording of Results.

Volume of pipette used = $25cm^3$ (Volume of BA_2 taken for each titration).

Burette readings;

Number of titrations	Rough	1	2	3
Final burette reading (cm^3)	20.1	40.40	20.00	40.20
Initial burette reading (cm^3)	0.00	20.40	0.00	20.00
Volume of titrant (cm^3)	20.1	20.00	20.00	20.20

Values used to calculate volume of BA_1, 20.0, 20.0,20.1

Average volume of BA_1 used = (20.0 + 20.0 + 20.1)/3 = $20.0cm^3$

Questions:

(a) Calculate

i) The number of moles of acid that reacted

ii) The number of moles of BA_2 that reacted

iii) The molarity of BA_2

(b) Find the number of molecules of water of crystallization (n) in $Na_2CO_3\ nH_2O$

Solution:

In the above experiment, BA_1 is acting as a standard solution because its concentration is already known in moles per lite. We are going to use this solution to determine the concentration of BA_2

Note: No titration calculations can be done unless the correct chemical equation is known. You must correctly write down the equation of reaction and ensure that it perfectly balances. The method used in this book involves working from first principles.,

i) $1000cm^3$ of Acid (BA_1) contain 0.1 moles this means that,

$1cm^3$ of acid contains $\left(\frac{0.1}{1000}\right) x\ 20 = 0.002\ moles$

ii) From the reaction ration, $CO_3^{2-}{}_{(aq)} + 2H+ = 1:2$

Therefore, number of moles of carbonate $= \frac{1}{2}$ moles of acid $=$

$$\frac{0.1}{1000} x\ 20\ x \frac{1}{2} = \frac{0.002}{2} = 0.001\ moles$$

III) $25\ cm^3$ of BA_2 contain 0.001 moles

$1cm^3$ of BA_2 contains $\frac{0.001}{25}$

$1000\ cm^3$ of BA_2 contain $\left(\frac{0.001}{25}\right) x\ 1000 = 0.04\ M$

Therefore Molarity of BA_2 solution is 0.04M

IV) Since Molarity $= \left(\frac{concentration\ in\ gms/liter}{Relative\ Formula\ Mass}\right)$

Relative Formula Mass = $\dfrac{concentration(g\ per\ l)}{molarity}$

R.F.M $= \dfrac{11.4}{0.04} = 285g$

Do you remember that BA2 was made by dissolving 11.4 g of $Na_2CO_3\ nH_2O$ in 1 liter of water?

But, R. F. M of $Na_2CO_3\ nH_2O$ = $(23\times2)+ 12 +(16\times3) +(1\times2 +16)n$

This means; $(23\times2)+ 12 +(16\times3) +(1\times2 +16)n = 285$

$$106 + 18n = 285$$
$$18n = 285-106$$
$$18n = 179$$

$$\frac{18n}{18} = \frac{179}{18}$$

$$\underline{n = 9.9 = 10.}$$

Therefore the number of molecules of water of crystallization (n) is 10 and the formula of the substance becomes **$Na_2CO_3\ 10\ H_2O$**

3.4.2 Experiment 2

TO DETERMINE THE BASICITY OF AN ACID (H_xM)

Provided:

- 0.5M Sodium Hydroxide solution (BA_1)
- Phenolphthalein indicator
- Clamp and stands
- 3 conical flasks
- 50cm^3 Burette
- BA_2 solution
- 25cm^3 pipette

Procedure

Rinse and fill the burette with solution BA_2, which is a 0.25M acid solution of formula H_xM. Pipette out BA_1 into a conical flask. Add 2-3 drops of phenolphthalein indicator. Titrate with BA_2 from the burette until you reach an end point. Repeat the above procedure until you get 3 consistent results.

Recording of results

Volume of pipette used = 25cms (volume of BA_1)

Burette readings

Number of titrations	Rough	1	2	3
Final burette reading (cm^3)	16.30	32.8	16.40	33.00
Initial burette reading (cm^3)	0.00	16.40	0.00	16.50
Volume of titrant (cm^3)	16.30	16.4	16.40	16.50

Values used to calculate volume of $BA_2 =$, 16.4, 16.4, 16.5

Average volume of BA_2 used = (16.4 + 16.4 + 16.5)/3 = 16.4 cm^3

Questions

a)

i) Write the molecular equation of the reaction

ii) Write the ionic equation of the reaction

b) Calculate the basicity of the Acid

Solution:

a) i)

Molecular equation of reaction is;

$$xNaOH_{(aq)} + H_xM_{(aq)} \rightarrow Na_xM_{(aq)} + xH_2O_{(l)}$$

ii) Ionically

$$xOH^-_{(aq)} + xH^+_{(aq)} \rightarrow xH_2O_{(l)}$$

b) Moles of (BA_1) Sodium Hydroxide that reacted

$1000cm^3$ contain 0.5 moles

$1\ cm^3$ contain $\dfrac{0.5}{1000}$

$25cm^3$ contain $\dfrac{0.5}{1000}x25$

Moles of BA_1 (Sodium Hydroxide = <u>0.0125</u> moles

From the above reaction, a reaction ratio of base to acid can be established as x:1.

Therefore moles of acid that reacted with $25cm^3$ of base $= \frac{1}{x} x \left(\frac{0.5}{1000} x25\right)$ ....A

But moles of acid that reacted in 16.4 cm^3 $= \frac{0.25}{1000} x16.4$ ....B

Do you remember that the molarity of (BA_2) acid was 0.25M?

Therefore if we can compare or equate the two equations A and B,

$$\frac{1}{x} x \left(\frac{0.5}{1000} x25\right) = \frac{0.25}{1000} x16.4$$

$$\frac{12.5}{1000X} = \frac{4.1}{1000}$$

$$\frac{4100X}{4100} = \frac{12500}{4100}$$

X= 3.05 = 3

The basicity of the acid is therefore 3, (H_3M). We can therefore say that this is a tri basic Acid since x = 3. The basicity of an acid refers to the number of hydrogen atoms in one molecule of that acid that are replaceable by a metal.

3.4.3 Experiment 3

TO DETERMINE THE RELATIVE ATOMIC MASS OF A METAL (X) IN A METAL HYGROGEN SULPHATE (XHSO$_4$)

Provided:

- 0.1M Sodium Hydroxide solution (BA$_2$)
- Phenolphthalein indicator
- 3 conical flasks
- Clamps and stands
- 3.4g of Metal hydrogen sulphate
- 20cm^3 pipette

Procedure:

Accurately weigh 3.4 g of a metal hydrogen sulphate (XHSO$_4$) provided and dissolve it in water in 250cm^3 volumetric flask and make the solution up to the mark and label it BA$_1$ Pipette 20 cm^3 of BA$_1$ into a conical flask, add 1-2 drops of phenolphthalein indicator.

Titrate with BA$_2$ from the burette, until you reach an end point. Repeat the above procedure until you get three consistent results.

Recording of results

Volume of pipette used = 20cm^3

Burette readings

Number of titrations	Rough	1	2	3
Final burette reading (cm^3)	19.5	40.3	42.55	20.10
Initial burette reading (cm^3)	0.00	20.20	22.50	0.00
Volume of titrant (cm^3)	19.50	20.10	20.05	20.10

Values used to calculate volume of BA_2 20.10, 20.05, 20.10

Average volume of BA_2 used = (20.10 + 20.05 + 20.10)/ 3 = 20.1 cm^3

Questions

a)

i) Write the equation of the reaction

ii) Calculate the Molarity of $XHSO_4$ (BA_1)

(b) Calculate the relative formula mass of the metal X in $XHSO_4$

Solution:

a) (i) Equation of the reaction is:

$$2XHSO_{4\ (aq)} + 2NaOH_{\ (aq)} \rightarrow X_2SO_{4(aq)} + Na_2SO_{4(aq)} + 2H_2O_{\ (l)}$$

(II) From the equation of the reaction $XHSO_4$: $NaOH$ gives a reaction ratio of 1:1

Therefore Moles of BA_1 = Moles of BA_2

Moles of BA_2 (Sodium hydroxide) that reacted will be got as follows;

1000 cm^3 contain 0.1 moles

1cm^3 contains $\frac{0.1}{1000}$

20.06 cm^3 contain $\frac{0.1}{1000} x20.06 = 0.002\ moles$

From the reaction ratio, $BA_1 : BA_2$ = 1:1

But moles of BA_1 (metal hydrogen sulphate) = 0.002 moles

Therefore to get the Molarity of BA_1 we shall say,

$20cm^3$ contain 0.002moles

$1cm^3$ will contain $\frac{0.002}{20}$

$1000cm^3$ will contain $\frac{0.002}{20} x\ 1000 = 0.1$

Therefore Molarity of BA_1 (metal hydrogen sulphate) = 0.1M

(b)

$250\ cm^3$ of the solution of $XHSO_4$ contain 3.4g of $XHSO_4$

$1\ cm^3$ of this solution contains $\frac{3.4}{250}$

$1000cm^3$ will contain $\frac{3.4}{250} x\ 1000 = 13.6g$

This means that the concentration of $XHSO_4$ in 1 liter = 13.6g

But we know the molarity of BA_1 as 0.1 and since;

Molarity = $(\frac{concentration\ in\ gms/liter}{Relative\ Formula\ Mass})$

$0.1 = \frac{13.6}{x+1+32+(16x4)}$

$0.1 = \frac{13.6}{x+97}$

$0.1 = (x+97) = 13.6$

$0.1x + 9.7 = 13.6$

$0.1x = 13.6-9.7$

$0.1x = 3.9$

$\frac{0.1x}{0.1} = \frac{3.9}{0.1}$

x = 39.

Therefore the relative atomic mass of the metal X = 39. You can use the table given in the appendix 1 to determine which element is X. (Did you find out that it is potassium?)

3.4..4 Experiment 4

TO DETERMINE THE PERCENTAGE IMPURITY IN A SAMPLE OF IMPURE SODIUM HYDROXIDE

Provided:

- 0.1M HCl (BA_2)
- Phenolphthalein indicator
- 3 conical flasks
- Clamps and stands
- 25cm^3 pipette
- BA_1 made by dissolving 5g of impure Sodium Hydroxide in one liter of water.

Target; you are required to calculate the percentage impurity of Sodium hydroxide

Procedure:

Pipette 25 cm^3 of BA_1 into a conical flask and add 1-2 drops of phenolphthalein indicator. Add BA_2 from the burette until you reach the end point. Read and record your results in the table provided below. Repeat the above procedure until you get 3 consistent results

Recording of results

Volume of pipette used = 25cm^3

Burette readings

Number of titrations	Rough	1	2	3
Final burette reading (cm^3)	24.40	24.50	49.10	24.50
Initial burette reading (cm^3)	0.00	0.00	24.60	0.00
Volume of titrant (cm^3)	24.40	24.50	24.50	24.50

Values used to calculate volume of BA_2 24.50, 24.50,24.50

Average volume of BA_2 used = (24.50 + 24.50 + 24.5)/3 = 24.50 cm^3

Questions:

    a i) write the equation of the reaction that took place between BA_1 and BA_2

      ii) Calculate the Molarity of BA_1 sodium hydroxide

    b) Determine the percentage impurity of BA_1

Solution

    (a) I) Equation of reaction between BA_1 and BA_2 is

      $HCl_{(aq)} + NaOH_{(aq)} \rightarrow NaCl_{(aq)} + H_2O_{(l)}$

      Reaction ratio of acid to alkali = 1:1

(ii) To get Molarity of BA_1 we need to first get the number of moles of BA_1 (standard solution) and use it to get the number of moles of BA_2 hence its Molarity.

Therefore since we know the molarity of BA_2 as 0.1, we shall say,

1000cm^3 of BA_2 *contain 0.1moles*

$1cm^3$ *contain* $\dfrac{0.1}{1000}$

$24.5 cm^3$ $contain$ $\frac{0.1}{1000}x24.5 = 0.00245$ $moles$

but from the reaction ratio of acid to base, (BA₁ to BA₂) = 1:1 . This means, number of moles of BA_2 = number of moles of BA_1 0.00245 moles

Therefore to get molarity of BA_1, we shall say

25 cm^3 of BA_1 contain 0.0025

1cm^3 of BA_1 contains $\frac{0.00245}{25}$

1000cm^3 of BA_1 contain $\frac{0.00245}{25}x1000 = 0.098$

Molarity of BA_1 = 0.098 M

(b) To get percentage impurity, we need to first get the concentration of NaOH in g/litre

Since Molarity $= \dfrac{concentration\ in\ gms/liter}{Relative\ Formula\ Mass}$

$0.098 = \dfrac{concentration\ in\ gms/liter}{(23+16+1}$

$0.098 = \dfrac{concentration\ in\ gms/liter}{40}$

Concentration (of Sodium hydroxide) = 3.92

Mass of impurity in 1 liter of solution = 5-3.92 = 1.08

Percentage impurity =

$$\frac{mass\ of\ impurity\ in\ 1l\ of\ solution}{original\ mass\ of\ impure\ sample} \times 100$$

$$= \frac{1.08}{5} x\ 100 = 21.6\%$$

Percentage impurity of BA_1 = **21.6%**

3.4.5 Experiment 5

TO DETERMINE THE RELATIVE MOLECULAR MASS OF AN ACID H_2M

Provided:

- 0.12 M NaOH (BA_1)
- Phenolphthalein indicator
- 3 conical flasks
- Clamps and stands
- $25cm^3$ pipette
- BA_2 made by dissolving 6.3g of a dibasic acid (H_2M) in 1L water

Procedure:

Pipette $25cm^3$ of BA_1 into a conical flask and add 3 drops of phenolphthalein indicator. Titrate this solution with BA_2 from the burette until you reach the end point. Repeat the above procedure until you get three consistent results. Record your results in the table below.

Recording of results

Volume of pipette used = $25cm^3$ (Volume of BA_1)

Burette readings

Number of titrations	Rough	1	2	3
Final burette reading (cm^3)	29.50	29.90	30.00	30.00
Initial burette reading (cm^3)	0.00	0.00	0.00	0.00
Volume of titrant (cm^3)	29.50	29.90	30.00	30.00

Values used to calculate volume of BA_2 29.9, 30.0, 30.0,

Average volume of BA_2 used = (29.9 +30.0+ 30.0)/3 = 30.0 cm^3

Questions

(a) Calculate the Molarity of the Acid solution (BA_2)

(b) Calculate the relative molecular mass of the Acid (H_2M)

Solution:

It is very important to first write the equation of reaction before proceeding with titration calculations, knowing the reaction ratio helps you to relate number of moles reacting.

(a) The equation of reaction is

$$H_2X_{(aq)} + 2NaOH_{(aq)} \rightarrow Na_2X_{(aq)} + 2 H_2O_{(l)}$$

$1000 \ cm^3$ of BA_1 contain 0.12 moles

$1 \ cm^3$ contains $\dfrac{0.12}{1000}$

$25 cm^3$ contains $\dfrac{0.12}{1000} x \ 25$

From the equation of reaction, the reaction ratio of Acid to Alkali is 1:2. This means that, the number of moles of Acid that reacted = ½ of number of moles of alkalis.

$$= ½ \ x \ \dfrac{0.12}{1000} x \ 25 \ = \underline{0.0015 \ moles}$$

So to get Molarity of BA_2, (Acid solution), we shall say;

30 cm^3 of BA_2 contain 0.0015 moles

1 cm^3 of BA_2 contains $\dfrac{0.0015}{30}$

1000 cm^3 of BA_2 will contain $\dfrac{0.0015}{30} x1000 = 0.05$

Therefore, the Molarity of the Acid solution BA_2 is 0.05M

(b) Since Molarity $= \dfrac{concentration \ in \ grams/liter}{Relative \ Formula \ Mass \ (RFM)}$

$\text{R.F.M} = \dfrac{concentration \ in \ grams/liter}{Malarity}$

But concentration in g/l = 6.3

$\text{RMM} = \dfrac{6.3}{0.05} = 126$

The relative molecular mass of the Acid (H_2M) = 126g

3.4.6. Experiment 6

TO DETERMINE THE RELATIVE ATOMIC MASS OF THE METAL (M) IN A METAL CARBONATE (M_2CO_3)

Provided:

- 0.1 M HCl (BA_1)
- Phenolphthalein indicator
- 3 conical flasks
- Clamps and stands
- $25cm^3$ pipette
- BA_2 made by dissolving 5.75 g of a metal carbonate (M_2CO_3) in 1L of water

Aim: You are required to determine the relative Atomic Mass of M.

Hydrochloric Acid reacts with M_2CO_3 according to the mole ratio 2:1

Procedure:

Pipette $25cm^3$ of BA_2 into a conical flask and titrate it with BA_1 using phenolphthalein indicator. Repeat the titration until you obtain consistent results, record your results in the table below.

Recording of results

Volume of pipette used = $25cm^3$

Burette readings

Number of titrations	Rough	1	2	3
Final burette reading (cm^3)	27.20	27.10	27.00	47.10
Initial burette reading (cm^3)	0.00	0.00	0.00	20.00
Volume of titrant (cm^3)	27.20	27.10	27.00	27.10

Values used to calculate volume of BA_1 = 27.10, 27.00, 27.10

Average volume of BA_1 used = (27.1 + 27.0+ 27.1)/3 = 27.10 cm^3

Questions

(a) Write the ionic equation for the reaction between Hydrochloric Acid and M_2CO_3

(b) Calculate

 i) The number of moles of BA_1 that reacted

ii) The Molarity of BA_2

iii) The Relative Atomic Mass of M (C = 12, O = 16)

Solution

Note: In the above question BA_1 is acting as a standard solution.

(a) $2H^+_{(aq)} + CO_3^{2-}_{(aq)} \rightarrow H_2O_{(l)} + CO_{2\,(g)}$

(b)

 i) 1000 cm^3 of BA_1 contain 0.1 moles

   1cm^3 contain $\frac{0.1}{1000}$ $moles$

   27.1 cm^3 of BA_1 will contain $\frac{0.1}{1000}$ $x27.1 = 0.00271\ moles$

ii) From the reaction ration of Acid to alkali 2:1, this means that, moles of alkali $= \frac{1}{2}$ moles of acid.

$$= \frac{1}{2} x 0.00271 = 0.00135 \ moles$$

Therefore,

25cm^3 of BA$_2$ Contain 0.00135 moles

1cm^3 contains $\frac{0.00135}{25}$

1000cm^3 contain $\frac{0.00135}{25}$x 1000 = <u>0.0542M</u>

iii) Since Molarity $= \dfrac{concentration \ in \ grams/liter}{Relative \ Formula \ Mass \ (RFM)}$

$$0.0542 = \frac{5.75}{2m+12+(16x3)}$$

$$0.0542 = \frac{5.75}{2m+60}$$

0.0542(2m+60) =5.75

0.108m +3.252 = 5.75

0.108m = 5.75-3.252.

$$\frac{0.108m}{0.108} = \frac{2.498}{0.108}$$

M = 23.1 = 23

The Relative Atomic Mass of M = 23. You can check in appendix 1 of this book and find out which metal M should be.

3.4.7 Experiment 7

TO DETERMINE THE STOICHOMETRIC RATIO FOR THE REACTION BETWEEN ACID AND ALKALI

Provided:

- 0.2 M of acid (BA_2)
- 0.3M NaOH solution (BA_1)
- Methyl orange indicator
- 3 conical flasks
- Clamps and stands
- 20cm^3 pipette

Procedure:

Pipette 20cm^3 of BA_1 into a conical flask and titrate it with BA_2 using Methyl orange indicator. Repeat the titration until you obtain consistent results, record your results in the table below.

Recording of results

Volume of pipette used = 20cm^3

Burette readings

Number of titrations	Rough	1	2	3
Final burette reading (cm^3)	15.30	31.90	46.90	15.00
Initial burette reading (cm^3)	0.00	16.90	31.90	00.00
Volume of titrant (cm^3)	15.30	15.00	15.00	15.00

Values used to calculate volume of BA_2 15.0, 15.0, 15.0

Average volume of BA_2 used = (15. + 15+ 15)/3 = <u>15.0 cm^3</u>

Questions

(a) Calculate

    i)   The number of moles of BA2 that reacted

    ii)   The number of moles of BA1 that reacted

    iii)   The relative Atomic Mass of M (C=12, 0= 16)

(b) Calculate the molar ratio of the acid to alkali for the reaction

Solution

(a) **i)** 1000cm^3 of BA_2 contain 0.2moles

    1 cm^3 will contain $\dfrac{0.2}{1000}$

    15 cm^3 will contain $\dfrac{0.2}{1000} x\ 15 = 0.003\ moles$

    ii) Moles of BA_1 that reacted

    1000cm^3 of BA_1 contain 0.3 moles

    1cm^3 will contain $\dfrac{0.3}{1000}$

    20 cm^3 contain $\dfrac{0.3}{1000}$x 20 = 0.006 moles

(b) Molar ratio of Acid to Alkali = 0.003:0.006

                <u>= 1:2</u>

1mole of Acid reacts with 2 moles of Alkali

3.4.8 Experiment 8

TO IDENTIFY "X" IN "KHX"

Provided:

- 0.1 M of H_2SO_4 (BA$_2$)
- Solution (BA$_1$) containing 10.0g/l of an acid salt (KHX)
- Phenolphthalein indicator
- 3 conical flasks
- Clamps and stands
- 25cm^3 pipette

Note: X is a radical; it can be a carbonate, or sulphate.

Procedure:

Pipette 25cm^3 of BA$_1$ into a conical flask add 2-3 drops of phenolphthalein indicator and titrate it with BA$_2$ from the burette until end point is reached. Repeat the titration until you obtain consistent results, record your results in the table below.

Recording of results

Volume of pipette used = 25.00cm^3 (Volume of BA$_1$ used)

Burette readings

Number of titrations	Rough	1	2	3
Final burette reading (cm^3)	12.6	25.00	27.00	39.5
Initial burette reading (cm^3)	0.00	12.50	14.50	27.00
Volume of titrant (cm^3)	12.60	12.50	12.50	12.50

Values used to calculate volume of BA_2 = 12.5, 12.5, 12.5

Average volume of BA_2 used = (12.5. + 12.5+ 12.5)/3 = <u>12.5 cm^3</u>

Questions

(a) Calculate the Molarity of BA_1 (KHX)

(b) Calculate the Relative Formula mass of KHX

(c) Identify "X"

Solution

As already seen before, it is very important to first write the equation of reaction before proceeding with calculations.

The equation of this reaction is therefore,

$$2KHCO_{3\ (aq)} + H_2SO_{4\ (aq)} \rightarrow K_2SO_{4\ (aq)} + 2H_2O_{\ (l)} + 2CO_{2\ (g)}$$

I have used $KHCO_3$ instead of $KHSO_4$ because $KHSO_4$ is an acidic salt and cannot react with an acid. The only possible reaction therefore is when X is a CO_3.

(a) 1000cm^3 of BA_2 contain 0.1 moles

1cm^3 of BA_2 contain $\frac{0.1}{1000}$

12.5cm^3 of BA_2 contain $\frac{0.1}{1000}$x 12.5

From the above equation of reaction, 1 mole of BA_2 requires 2 moles of KHX. This means that, moles of BA_1 that reacted = 2x moles of BA_2

$$= 2x\frac{0.1}{1000}x\ 12.5 = \underline{0.0025moles}$$

These moles are in 25cm^3, we shall therefore say that,

25cm^3 of BA$_1$ contain 0.0025moles

1cm^3 contains $\frac{0.0025}{25}$

1000cm^3 of BA$_1$ contain $\frac{0.0025}{25} x\ 1000 = 0.1\ moles$

Therefore Molarity of BA$_1$ = $\underline{0.1M}$

(b) R.F.M = $\dfrac{concentration\ in\ grams/liter}{Molarity}$

but concentration = 10.0g

R.F.M = $\frac{10.0}{0.1} = 100g$

Relative Formula Mass of KHX is $\underline{100g}$

(c) The formula given is KHX. To get X, we shall say; (K=39, H=1)

39 +1 +X = 100

40+X   = 100

X   =100-40

X   = 60

We can confirm that X is a carbonate because,

(i)   The solution of KHX reacted with Sulphuric acid

(ii)   The RFM of X = 60, (CO_3 = 12+48 = 60)

(iii)   The solution is basic to Methyl Orange indicator

3.4.9 Experiment 9

TO DETERMINE THE RELATIVE FORMULA MASS OF 'MCO₃' AND ATOMIC MASS OF 'M'

Provided:

- 0.1 M NaOH
- 1M HCl
- Unknown insoluble metal carbonate (MCO_3)
- Methyl Orange indicator
- 250 cm^3 Volumetric flask
- 3 conical flasks
- Clamps and stands
- 25cm^3 pipette

Procedure:

Accurately measure 50cm^3 of 1M HCl and pour it into a 250cm^3 volumetric flask. Accurately weigh 1.5g of MCO_3 and transfer it into a 250cm^3 volumetric flask containing 50cm^3 of Hydrochloric Acid. Wash the container thoroughly well with distilled water and add the washing into the volumetric flask and shake well. After effervescence has stopped, make the solution up to the mark with distilled water and shake well.

Pipette 25cm^3 of this solution into a conical flask. Add 2-3 drops of Methyl Orange indicator. Add Sodium Hydroxide solution from the burette, titrate it against that in the conical flask until end point is reached. Record your readings in the table provided.

Recording of results

Volume of pipette used = 25.00cm^3

Mass of MCO_3 = 1.5g

Burette readings

Number of titrations	Rough	1	2	3
Final burette reading (cm^3)	20.20	20.00	42.20	20.00
Initial burette reading (cm^3)	0.00	0.00	22.20	0.00
Volume of titrant (cm^3)	20.20	20.00	20.00	20.00

Values used to calculate volume of NaOH, = 20.00, 20.00, 20.00

Average volume of NaOH used = (20.0+20.0+20.0)/3 = 20.0 cm^3

Questions:

(a) Calculate

   (i)   The number of moles of NaOH that reacted with 25cm^3 of excess acid

   (ii)   The number of moles of HCl present in 25cm^3 of excess acid

   (iii)   The number of moles of HCl that reacted with MCO_3

   (iv)   The Relative Molar Mass of MCO_3

   (v)   The relative Atomic Mass of X, given that (C= 12, O=16)

Solution

(a) Equations of reactions occurring are

$$MCO_3 {}_{(s)} + 2HCl {}_{(aq)} \rightarrow MCl_2 {}_{(aq)} + CO_{2(g)} + H_2O_{(l)}$$

$$HCl {}_{(aq)} + NaOH_{(aq)} \rightarrow NaCl_{(aq)} + H_2O {}_{(l)}$$

(i)   Number of moles of NaOH that reacted

$1000cm^3$ of NaOH contain 0.1moles

$1cm^3$ contains $\dfrac{0.1}{1000}$

20cm3 of NaOH contain $\dfrac{0.1}{1000} \times 20 = 0.002 moles$

(ii)   From the equation given above, 1 mole of NaOH reacts with 1mole of HCl. This means that, moles of NaOH = moles of HCl = 0.002moles. But this number of moles is in $25cm^3$ of excess Acid.

(iii) $25cm^3$ of excess acid contain 0.002moles

$250cm^3$ of excess acid contain $\dfrac{0.002}{25} x250 = 0.02 moles$

To get the number of moles of HCl that reacted with MCO_3, we shall subtract moles of excess acid from moles of acid in the original 50cm3 of 1M HCl.

But moles of acid in the original $50cm^3$ of 1M HCl can be got as below

$1000cm^3$ contain 1.0 mole.

$1cm^3$ contain $\frac{1}{1000}$

$50cm^3$ contain $\frac{1}{1000} x50 = 0.05 moles$

Moles of HCl that reacted with MCO_3= Moles in original $50cm^3$ - moles of excess acid

= 0.05-0.02

= 0.03moles

Therefore number of moles of HCl that reacted with MCO_3 = 0.03 moles

(iv)   From the equations of reaction given above,

1 mole of MCO_3 reacts with 2 moles of HCl

This means that, the number of moles of MCO_3 that reacted with HCL =

$$\frac{1}{2}X0.03 = 0.015\ moles$$

These are the moles of MCO_3 corresponding to the mass of MCO_3 that is 1.5g

We shall therefore say that,

Number of moles = $\frac{Mass\ provided}{Molar\ Mass\ (RFM)}$

$0.015\ = \frac{1.5}{RFM}$

$$RFM = \frac{1.5}{0.015} = 100$$

The Relative Molar Mass of MCO_3 = 100g

(V) Let M represent the Atomic Mass in MCO_3 Such that,

M+12 +(16x3) = 100

M +12+48 = 100

M+60 =100

M= 100-60

M = 40

The Atomic Mass of M in MCO_3 is therefore 40. This compound is likely to be $CaCO_3$ (use the appendix 1 of this book to identify M)

3.4.10 Experiment 10

TO DETERMINE THE PERCENTAGE PURITY OF A SUBSTANCE

Provided:

- 0.1 M HCl
- Methyl Orange indicator
- 250 cm^3 Volumetric flask
- 3 conical flasks
- Clamps and stands
- 25cm^3 pipette
- Sample of impure carbonate

Procedure:

Accurately weight 1.7g of impure sample of carbonate, put it into a 250cm^3 volumetric flask. Wash the container used with distilled water and add the washing into the volumetric flask. Add about 50cm^3 of distilled water and shake well until the entire solid dissolves. Make the solution up to the mark and shake.

Pipette 25cm^3 of this solution into a conical flask. Add 2-3 drops of Methyl Orange indicator. Add Hydrochloric Acid from the burette until the end point is reached. Repeat the titration until you get consistent results. Record your readings in the table provided.

Recording of results

Volume of pipette used = 25.00cm^3

Mass of impure carbonate= 1.7g

Burette readings

Number of titrations	Rough	1	2	3
Final burette reading (cm^3)	20.60	20.20	40.30	20.20
Initial burette reading (cm^3)	0.00	0.00	20.20	0.00
Volume of titrant (cm^3)	20.60	20.20	20.10	20.20

Values used to calculate volume of HCl, = 20.00, 20.10, 20.20

Average volume of HCl used = (20.0+20.1+20.2)/3 = <u>20.2 cm^3</u>

Questions:

(a) Calculate the Molarity of potassium carbonate in solution

(b) Calculate the percentage purity of the original sample (H=1, O= 16, K=39, C=12)

Solution.

(a). You must always write the correct equation of reaction before proceeding with calculations.

K_2CO_3 $_{(aq)}$ + 2HCl $_{(aq)}$ → KCl $_{(aq)}$ + H_2O $_{(l)}$ + CO_2 $_{(g)}$

HCl is acting as a standard solution because it's molarity is already known.

1000cm^3 contain 0.1 moles

1cm^3 contains $\frac{0.1}{1000}$

20.2cm^3 contain $\frac{0.1}{1000} x20.2 = 0.00202\ moles$

From the reaction ratio, 2moles of HCl react with 1 mole of K_2CO_3 This means that, number of moles of K_2CO_3 = ½ moles of HCl.

$$\frac{1}{2} x \frac{0.1}{1000} \text{ x } 20.2 = \underline{0.00101 \text{ moles}}$$

We shall therefore say;

$25cm^3$ of K_2CO_3 contain 0.00101 moles

$1cm^3$ of K_2CO_3 contains $\frac{0.00101}{25}$

$1000cm^3$ of K_2CO_3 contain $\frac{0.00101}{25} x 1000 = 0.0404\ moles$

Therefore the molarity of K_2CO_3 is $\underline{0.0404 \text{ M}}$

(b) To get percentage purity, we shall say that,

$$\% \text{ Purity } = \frac{Mass\ of\ pure\ \ K2CO3\ in\ grams\ per\ liter}{mass\ of\ original\ sample\ in\ 1\ liter}$$

But mass of pure carbonate in 1 litre = Molarity x Molar mass

Molar mass of K_2CO_3 = (39x2) + (16x3) + 12 = 138

= 138x0.0404 = 5.5752 g/l

Mass of original sample in 250 cm^3 = 1.7

Mass of original sample in 1 Litere (1000cm^3) $= \frac{1.7}{250} x\ 1000 = 6.8\ g\ in\ 1\ litre$

% purity $= \frac{5.5752}{6.8} x 100 = 81.9\%$

Therefore percentage purity of the original sample $= 82\%$.

3.4.11 Experiment 11

TO DETERMINE THE RATE OF A CHEMICAL REACTION BY PRECIPITATION OF SULPHUR USING DILUTE HYDROCHLORIC ACID ON SODIUM THIOSULPHATE SOLUTION

You are provided with BA_1, which is a solution made by dissolving 20g of sodium thiosulphate in 1 litre of water.

You are also provided with BA_2, which is a solution of 1.0M, Hydrochloric Acid, and a stop clock

BA_2 reacts with BA_1 according to the following equation:

$$Na_2S_2O_3 \text{ (aq)} + 2HCl_{(aq)} \rightarrow S_{(s)} + 2NaCl \text{ (aq)} + H_2O_{(l)} + SO_{2 \text{ (g)}}$$

Procedure:

Measure 50 cm^3 of BA_1 and place it in a conical flask. Place the flask and its contents on a sheet of paper with a cross-marked on it. Add 5 cm^3 of BA_2 and start the stop clock.

Note the time taken for the cross to disappear. Repeat the experiment by varying the concentration of BA_1 i.e measure 40cm^3, 30cm^3, 20cm^3, and 10cm^3 of BA_1 into a conical flask and make the volume 50cm^3 by adding 10cm^3, 20cm^3, 30cm^3, and 40cm^3 of water respectively and note the time taken for the cross to disappear in each case.

Table of results

Volume of BA_1 (cm^3)	50	40	30	20	10
Volume of water used cm^3	0	10	20	30	40
Concentration of BA_1 (cm^3)	1	0.8	0.6	0.4	0.2
Time (s)	30	50	65	85	140
1/t (s^{-1})	0.03	0.02	0.015	0.012	0.007

Note: The concentration per liter for each volume of BA_1 used was obtained as below,

For $50cm^3$ concentration $= \frac{20}{1000} x\ 50\ = 1gl^{-1}$

For $40\ cm^3$ concentration $= \frac{20}{1000} x\ 40\ = 0.8gl^{-1}$

For $30\ cm^3$ concentration $= \frac{20}{1000} x\ 30\ = 0.6gl^{-1}$

For $20cm^3$ concentration $= \frac{20}{1000} x\ 20\ = 0.4\ gl^{-1}$

For $10cm^3$ concentration $= \frac{20}{1000} x\ 10\ = 0.2\ gl^{-1}$

(20g is the mass of sodium thiosulphate in 1 liter of water.)

Questions

a) Plot a graph of

i) Concentration of BA_1 against time.

ii) 1/time (reciprocal of time) against concentration ratio of BA_1

b) Explain the nature of the two graphs obtained above

Concentration $\alpha \dfrac{1}{time}$

Rate of reaction $= \dfrac{1}{time}$

Note: The gradient of the curve in (i) gives the rate of reaction at any point.

The rate of reaction is the gradient of the reduction in concentration or pressure of the reactants. It is the rate at which the reactants disappear as the concentration of the products increases.

3.5 RELATED EXAMPLES

Example 1

You are provided with BA_1 which was made by dissolving 7.4g of an acid of molecular formula $H(CH_2)nCOOH$ in water to make one liter of solution.

BA_2 is a solution made by dissolving 4.96g of NaOH in 1liter of water.

AIM. You are required to determine the value of n in $H(CH_2)nCOOH$.

25Ccm^3 of BA_1 were titrated with BA_2 using phenolphthalein indicator and 20cm^3 of BA_2 were needed to reach the end point.

Questions

    (a) Calculate

        i)   The concentration of BA_1 in moles /liter

        ii)   The value of "n'

Solution

Equation of reaction

$$H(CH_2)nCOOH_{(aq)} + NaOH_{(aq)} \rightarrow H(CH_2)nCOONa_{(aq)} + H_2O_{(l)}$$

Molar mass of NaOH = 23 +16 + 1 = 40g.

Molarity of NaOH $= \frac{4.96}{40} = 0.124$

Moles of NaOH (BA_2) that reacted $= \frac{0.12}{1000} x20 = 0.0024$

But from the reaction ratio of Acid: Alkali = 1:1

Moles of BA_1 = Moles of BA_2

Therefore,

$25cm^3$ of BA_1 contain 0.0024moles

$1cm^3$ contains $\frac{0.0024}{25}$

$1000cm^3$ contain $\frac{0.0024}{25} x1000 = 0.096moles$

Therefore Molarity of BA_1 is 0.096 M. This is the concentration of BA_1 in grams per liter.

(ii). Molar mass of BA_1 =

$$Molar\ mass\ = \frac{Concentration\ in\ grams\ per liter}{Molarity}$$

$$Molar\ mass\ \ of\ BA_1 = \frac{7.4}{0.096} = 75g$$

We can therefore say that,

$H(CH_2)nCOOH = 1+12n+2n+12+16+16+1 = 75$

$14n + 46 = 75$

$14n + 46 - 46 = 75 - 46$

$14n = 29$

$\frac{14n}{14} = \frac{29}{14}$

$n = 2.07 = 2.$

The structural formula of BA_1 becomes **CH_3CH_2COOH**.

Example 2

You are provided with BA_1, which was made by dissolving 2.3625g of the acid of molecular formula $X-CH_2COOH$ in water to make $250cm3$ of solution. BA_2 is a solution made by dissolving 1g of sodium hydroxide pellets to make $250cm^3$ of solution.

$25cm^3$ of BA_1 was pipetted into a conical flask and titrated with $25.1cm^3$ of BA_2 using methyl orange indicator.

Questions.

(a) Calculate the molecular mass of the acid $X-CH_2\ COOH$

(b) Calculate the atomic weight of the element whose symbol is X

Hint. The equation of the reaction between the acid and sodium hydroxide is;

$$X-CH_2\ COOH_{(aq)} + NaOH_{(aq)} \rightarrow X-CH_2\ COONa_{(aq)} + H_2O_{(l)}$$

Solution

(a) We need to get the concentration of sodium hydroxide in $1000cm^3$ (1 Liter of solution) because $250cm^3$ contained only 1g, we shall therefore say that;

$250cm^3$ contain 1g of NaOH

$1cm^3$ contain $\frac{1}{250}$

$1000cm^3$ contain $\frac{1}{250}\ x\ 1000 = 4g$

This is the concentration in grams per liter.

We can now get the molarity of sodium hydroxide as follows,

$$\text{Molarity} = \frac{Concentration\ in\ grams\ perliter}{Relative\ Molecular\ Mass}$$

$$= \frac{4}{40}$$

Molarity of Sodium hydroxide is 0.1M

This means that,

1000cm^3 of Sodium hudroxide contain 0.1moles.

25.1cm^3 of this solution contain $\frac{0.1}{1000} \; x \; 25.1 = 0.00251 \; moles$

The number of moles of BA$_2$ that reacted = 0.00252

From the reaction ratio 1:1 according to the equation of reaction above, we shall say that the number of moles of BA$_1$ = moles of BA$_2$ = 0.00251 moles

Therefore,

25cm^3 of BA$_1$ contain 0.00251 moles

1cm^3 of BA1 contain $\frac{0.00251}{25}$

1000cm^3 of BA$_1$ contain $\frac{0.00252}{25} \; x \; 1000 = 0.1004 \; moles$

The molarity of the Acid solution BA$_1$ = 0.1004

But Molarity $= \dfrac{Concentration \; in \; grams \; perliter}{Relative \; Molecular \; Mass}$

Relative Molecular mass $= \dfrac{Concentration \; in \; grams \; perliter}{Molarity}$

But 250cm3 of BA$_1$ contain 2.3625g

$1cm^3$ of BA$_1$ contains $\frac{2.3625}{250}$

$1000cm^3$ of BA$_1$ contain $\frac{2.3625}{250} x1000 = 9.45\ g$

So relative formula mass $= \frac{9.45}{0.1004} = 94.12g$

$$Molecular\ formula\ mass = \underline{94.12\ g}$$

<u>(b)</u>

Since Molecular mass of the acid is 94.12g,

Using the molecular formula given, X-CH$_2$COOH

    X + (12+2+12+32+1) = 94.12

    X +59 = 94.12

    X =94.12-59

    <u>**X = 35.12**</u>

This is the atomic weight of X, which is equal to that of Chlorine. The acid is therefore ClCH$_2$COOH (Chloroethanoic Acid)

Example 3

A sample of pure Sodium Carbonate (Na_2CO_3) weighing 0.3542g is dissolved in 1Litre of water and titrated with a solution of Hydrochloric Acid (HCl) from the burette. A volume of 30.23cm^3 is required to reach the Methyl end point. The reaction being,

$$Na_2CO_3 \text{ (aq)} + 2HCl \text{ (aq)} \rightarrow 2NaCl \text{ (aq)} + H_2O \text{ (l)} + CO_2 \text{ (g)}$$

Question

Calculate, the molarity of the acid at equivalence point.

Solution

We need to first get the number of moles of Sodium Carbonate n 1 Liter of Solution

$$\text{Molarity} = \frac{Concentration\ in\ grams\ perliter}{Relative\ Formula\ Mass}$$

But R.F.M of Na_2CO_3 = (2x23) + (3x16) +12 = 106

Therefore Molarity $= \frac{0.3542}{106} = 0.0033M$

This means that, 1Litre of water contain 0.003moles of Na_2CO_3

From the reaction ratio Na_2CO_3 : 2HCl =1:2

Moles of HCl = 2xmoles of Carbonate at equivalence point)

30.23cm^3 contain 2(0.0033)moles

1cm^3 contain $\frac{0.0066}{30.23}$

1000cm^3 contains $\frac{0.0066}{30.23} x1000 = 0.2211moles$

At equivalence point, the Molarity of the acid is <u>0.2211M</u>

Example 4:

$100cm^3$ of concentrated Hydrochloric acid were diluted with distilled water to make 1liter of solution. $25cm^3$ of 0.5M Sodium Carbonate solution were pipetted out and $26.8cm^3$ of the acid solution were found to neutralize it using methyl orange indicator.

Question:

What is the concentration of the original acid in grams per liter?

Solution:

The equation of reaction is;

$$Na_2CO_{3\ (aq)} + 2HCl_{\ (aq)} \rightarrow 2NaCl_{\ (aq)} + H_2O_{\ (l)} + CO_{2\ (g)}$$

Ionic equation is,

$$2H^+_{\ (aq)} + CO_3^{2-}_{\ (aq)} \rightarrow CO_{2\ (g)} + H_2O_{\ (l)}$$

$1000cm^3$ of sodium Carbonate contains 0.5moles

$1cm^3$ contains $\frac{0.5}{1000}$

$25cm^3$ contain $\frac{0.5}{1000} x25$

Number of moles of Na_2CO_3 in $25cm^3$ is $\frac{0.5}{1000} x25$

From the equation of reaction above, the reaction ratio of Na_2CO_3 : HCl = 1:2, showing that, 1 mole of Na_2CO_3 reacts with 2 moles of HCl.

Therefore moles of HCl = 2x moles of Na_2CO_3 = $2 \times \frac{0.5}{1000} \times 25$ 0.025 moles.

To get Molarity of the acid, we shall say that,

$26.8cm^3$ of acid contains 0.025moles.

$1cm^3$ contains $\frac{0.025}{26.8}$

$1000cm^3$ of acid contains $\frac{0.025}{26.8} \times 1000 = 0.933 \; moles$

But 0.933 moles were in $100cm^3$ of concentrated HCl, meaning that, the concentrated acid contains

= 0.933x10 moles of the acid per liter

= 9.33moles

The mass of 1 mole of HCl = (1+35.5) = 36.5g

This means that, 1 Liter of concentrated HCl contains

9.33 x 36.5g of HCl = <u>340g.</u>

Therefore the concentration of the original HCl = <u>340 g/l</u>

Example 5.

0.05 M of solution of H_2SO_4 acid was titrated against an alkali of formula XOH. $25.0cm^3$ of alkali require $24.0cm^3$ of the acid for complete neutralization.

Questions:

(a) Write the equation of the reaction

(b) Calculate the number of moles of

    (i)   The acid used

    (ii)   Alkali in $25.0cm^3$ of solution

    (iii)   Alkali in $1000cm^3$ of solution

(c) If the alkali contains 3.0g in $800cm^3$ of solution. Calculate

    (i)   The concentration of the alkali in g/l of solution

    (ii)   Molar mass of the alkali

    (iii)   Relative atomic mass of X.

Solution

(a) Equation of reaction

$$H_2SO_{4\ (aq)} + 2XOH_{\ (aq)} \rightarrow X_2SO_{4\ (aq)} + 2H_2O_{\ (l)}$$

(b) (i)

    $1000cm^3$ of acid solution contain 0.05moles

    $1cm^3$ will contain $\frac{0.05}{1000}\ moles$

    $24.0\ cm^3$ of the acid solution contain $\frac{0.05}{1000} x\ 24 = 0.0012\ moles$ of H_2SO_4

    (II) From the reaction ratio, $XOH : H_2SO_4 = 2:1$

    This means that, the moles of XOH = 2x moles of acid

      = 2x 0.0012 = <u>0.0024 moles</u> of XOH

(III)

25.0cm^3 of alkali contain 0.0024moles

1cm^3 will contain $\frac{0.0024}{25}$

1000cm^3 of alkali contain $\frac{0.0024}{25} x1000 = 0.096M$

This is the molarity of XOH.

(C) (I)

If 800cm^3 contain 3.0g

1cm^3 contains $\frac{3}{800}$

1000cm^3 contain $\frac{3}{800} x1000 = 3.75g\ of$ XOH

The concentration of XOH = 3.75g/l

(ii) If 0.096 moles of XOH are contained in 3.75g

1 mole will contain $\frac{3.75}{0.096} = 39.0625g$

The molar mass of the alkali therefore is 39.062g

iii)   R.A.M = Formula mass of XOH = X+16+1

R.A. M = X+17

39.0625 = X+17

X= 39.0625 – 17

X = 22.0625

The relative atomic mass of X is 22.0625

Example 6

1.0g of metal carbonate XCO_3 was dissolved in 50cm^3 of IM Nitric Acid. The excess acid required 30.0cm^3 of 1 M potassium hydroxide for complete neutralization.

(a) Calculate the number of moles of the original acid

(b) Write the equation of reaction between potassium hydroxide and Nitric Acid.

(c) Calculate the number of moles of

   (i)   Potassium hydroxide that reacted with excess acid

   (ii)   Excess acid

   (iii)   Acid that reacted with XCO_3

(d) Write an equation for the reaction between XCO_3 and Nitric Acid

(e) Find the number of moles of XCO_3 that reacted with the acid

(f) Calculate the Molar mass of XCO_3 and hence the relative atomic mass of X.
   (C=12, 0=16)

Solution.

a)

1000cm^3 of solution contain 1 mole of HNO_3

50cm^3 of solution contain $\frac{1}{1000} x50 = 0.05\ moles$ of original acid

b)

$KOH_{(aq)} + HNO_{3(aq)} \rightarrow KNO_{3(aq)} + H_2O_{(l)}$

(c) i)

1000 cm^3 of solution contain 1 mole of KOH

30cm^3 of solution contain $\frac{1}{1000} x30 = 0.03 moles$ of KOH

ii) From the equation of reaction above, the reaction ration of Acid : Base = 1:1

Moles of excess acid = Moles of KOH = 0.03 moles

iii) Moles of HNO_3 acid that reacted with XCO_3

Original moles of HNO_3 – Excess moles of HNO_3

= 0.05 – 0.03

= 0.02 moles of HNO_3 that reacted with XCO_3.

(d) $XCO_{3(aq)} + 2HNO_{3\ (aq)} \rightarrow X(NO_3)_{2\ (aq)} + CO_{2\ (g)} + H_2O_{\ (l)}$

(e) Reaction ratio of XCO_3 : HNO_3 = 1:2

Moles of XCO_3 = ½ moles of HNO_3 = ½ x 0.02 = 0.01 moles of XCO_3

(f) 0.01 moles of XCO_3 are contained in 1g of XCO_3

1 mole of XCO_3 will be contained in $\frac{1}{0.01} = 100g$

To get the R.A.M of X in XCO3, we shall say that,

R.A.M = Formula mass

R.A.M = X+12+(16x3)

R.A.M = X+ 60

100 = X+60

X= 100-60

X = 40

Therefore the R.A.M of X is 40.

Using the table of atomic weights given in appendix 3 of this book, you will be able to identify element X.

Example 7

1.260g of crystal of a dibasic acid was dissolved in water to make 250cm^3 of solution. 20cm^3 of this solution required 16.0cm^3 of 0.1M of sodium hydroxide for complete neutralization.

Questions.

If the formula mass of the anhydrous acid is 90g, calculate;

    (a) The number of moles of Sodium Hydroxide used.

    (b) Number of moles of the acid in 20cm^3 of solution

    (c) Number of moles of acid in 25ocm^3 of solution

    (d) Molar mass of the acid (H =1, 0= 16)

Solution:

    (a) As already seen, we cannot proceed with calculation without first writing the equation of reaction.

$NaOH_{(aq)} + H_2X_{(aq)} \rightarrow Na_2X_{(aq)} + 2H_2O_{(l)}$

1000cm^3 of solution of NaOH contain 0.1moles

16.0cm^3 of this solution contain $\frac{0.1}{1000} x16 = 0.0016 \; moles$ of NaOH

    (b) To get moles of acid, we will use the reaction ratio between acid and base.

        $H_2X : 2NaOH = 1:2$

        This means that, moles of acid = ½ x moles of base.

        $= \frac{1}{2} x0.0016 = 0.0008 \; moles$ of H$_2$X in 20cm^3

(c) To get moles of acid in 250cm^3, we shall say that,

If 20cm^3 contain 0.0008 moles

1cm^3 will contain $\frac{0.0008}{20}$ $moles$

250cm^3 will contain $\frac{0.0008}{20} x250 = 0.01$ $moles$ of H$_2$X

(d) Molar mass of the acid

0.01 Moles of acid contain 1.2605g

1 mole contains $\frac{1.2605}{0.01} = 126.05$

Therefore, molar mass of the acid = 126.05g

Example 8:

A sample of a divalent metal Q, contaminated with its oxide was dissolved in 55cm^3 of 0.1M hydrochloric acid. 25.0cm^3 of hydrogen measured at s.t.p was evolved. 25cm^3 of 0.1M sodium hydroxide solution was required to neutralize the excess acid.

Questions:

Calculate the percentage of metal Q in the sample (1 mole of a gas at s.t.p occupies 22.4l or 22400cm^3)

Solution:

The equations of reaction occurring are as below;

(i) $Q_{(s)} + 2HCl_{(aq)} \rightarrow QCl_{2(aq)} + H_{2\,(g)}$

(ii) $QO_{(s)} + 2HCl_{(aq)} \rightarrow QCl_{2\,(aq)} + H_2O_{(l)}$

(iii) $NaOH_{(aq)} + HCl_{(aq)} \rightarrow NaCl_{(aq)} + H_2O_{(l)}$

To get the moles of hydrogen gas produced, we shall say that,

$22400 cm^3$ are occupied by 1 mole of hydrogen

$1 cm^3$ will be occupied by $\frac{1}{22400}$ $moles$

$25 cm^3$ will be occupied by $\frac{1}{22400} x25 = 0.001116\ moles$

From equation (i) above, moles of Q in the sample = moles of hydrogen gas produced = 0.001115 moles.

To get the number of moles of Hydrochloric acid, we shall say;

$1000 cm^3$ contain 0.1 moles

$1\ cm^3$ contains $\frac{0.1}{1000}$

$55 cm^3$ contain $\frac{0.1}{1000} x\ 55 = 0.0055\ moles$

Moles of NaOH that reacted with excess HCl will be ,

$1000 cm^3$ of solution of NaOH contain 0.1 moles

$1 cm^3$ contain $\frac{0.1}{1000}$

$25cm^3$ of solution will contain $\frac{0.1}{1000} x25 = 0.0025 \; moles$

Moles of HCl that reacted with base = 0.005-0.0025 = 0.0025moles.

But moles of HCl that reacted with metal Q = 2 x 0.001116 = 0.002232

Example 9

In an experiment to determine the percentage purity of a sample of Ammonium Chloride containing sodium Chloride as an impurity, the following were provided. 2M sodium hydroxide solution and 0.5M hydrochloric acid. $150cm^3$ of sodium hydroxide were measured and poured into a $250cm^3$ volumetric flask. 15.5g of an impure substance were weighed and added to the volumetric flask containing $150cm^3$ of sodium hydroxide. The contents of the flask were boiled for 4 minutes and the solution made up to $250cm^3$ with distilled water. $25cm^3$ of this solution required $22.4cm^3$ Of 0.5M Hydro Chloric Acid for complete neutralization using phenolphthalein indicator.

Questions:

Calculate

      (a) Number of moles of hydrochloric acid used.

      (b) Moles of sodium hydroxide in $250cm^3$ after dilution, which is the moles of excess NaOH in $250cm^3$.

      (c) Moles of Sodium hydroxide that reacted with Ammonium Chloride

      (d) Mass of Ammonium Chloride present in the sample

      (e) Percentage purity of the original sample.

      (Given that N=14, H =1, Cl = 35.5).

Solution.

The equation for the reaction is;

$NH_4Cl_{(s)} + NaOH_{(aq)} \rightarrow NaCl_{(aq)} + NH_{3(g)} + H_2O_{(l)}$ ....(i)

The ammonia gas was liberated after boiling the solution, it therefore does not take part in further reactions.

$NaOH_{(aq)} + HCl_{(aq)} \rightarrow NaCl_{(aq)} + H_2O_{(l)}$...(ii)

(a) 1000cm3 of HCl contain 0.5moles

22.4 cm3 contain $\frac{0.5}{1000} x22.4 = 0.0112\ moles$

(b) From equation (ii) above, 1 mole of NaOH neutralized 1 mole of HCl, reaction ratio = 1:1

This means that, moles of HCl = moles of NaOH = 0.0112 moles

These moles are in 25cm3.

25cm3 contain 0.0112 moles

250cm3 contain $\frac{0.0112}{25} x250 = 0.112$

These are the moles of excess Sodium hydroxide.

(c) The moles of NaOH that reacted with NH_4Cl

= moles of original 150cm^3 of 2M NaOH – moles of excess NaOH

But moles of original 150cm^3 of 2M NaOH =

$$\frac{150}{1000} X 2 = 0.3 moles$$

Moles of NaOH that reacted with NH_4Cl = 0.3-0.112 = <u>0.188 moles</u>.

(d) From the equation (i) above, 1 mole of NH_4Cl reacts with 1 mole of NaOH. Reaction ratio = 1:1.

This means that, moles of NH_4Cl = 0.188 moles

These moles are moles of NH_4Cl in the sample.

Molar mass of NH_4Cl = (14+4+35.5)

1 mole of NH_4Cl weighs 53.5g

0.188 moles of NH_4Cl weigh 53.5 x 0.188 = 10.058g

Hence mass of NH_4Cl in the sample = 10.058g

(e)　　percentage purity = $\dfrac{Mass\ of\ NH4Cl}{Mass\ of\ Sample} x100$

$$= \frac{10.058}{15.5} x100 = 64.9$$

Percentage purity of original sample = <u>64.9%</u>

Test Yourself

1. You are provided with BA_1, which was made by dissolving 2.52g of oxalic acid $(H_2C_2O_4nH_2O)$ in water to make $500cm^3$ of solution, BA_1 which is 0.05M sodium hydroxide solution and Phenolphthalein indicator.

Aim: You are required to determine the number of moles of water of crystallization (n) in one molecule of Oxalic Acid

Procedure:

Pipette $25.0cm^3$ of BA_1 into a conical flask. Add 3 drops of phenolphthalein indicator. Titrate this solution with BA_2 from the burette until the solution changes to pale pink. Repeat the procedure three more times to get consistent results. Record your results in the table below.

Table of results

Volume of pipette used = ...cm^3

Burette readings

Number of titrations	Rough	1	2	3
Final burette reading (cm^3)	39.90	47.90	48.70	39.30
Initial burette reading (cm^3)	0.00	08.20	09.00	0.00
Volume of titrant (cm^3)	39.90	39.70	39.70	39.30

Find,

    1. Volume of BA_2 used to calculate the average titer ....cm^3

    2. Average volume of BA_2 used...cm^3

3. Write an equation for the reaction that took place between BA_1 and BA_2

4. (a) Calculate the number of moles of

  i) BA_1 used

  ii) BA_2 used in 25 cm^3 of solution

  iii) BA_2 in 1000cm^3 of solution

  (b) Calculate the molar mass of $H_2C_2O_4.nH_2O$

  (C) calculate the value of 'n' (C= 12, H=1, O=16)

2. You are provided with BA_1, which is 1M HCl, and BA_2, which is 2M HCl, substance P that is magnesium powder. BA_1 reacts with P according to the reaction below.

$$2HCl_{(aq)} + Mg_{(s)} \rightarrow MgCl_{2\,(aq)} + H_{2\,(g)}$$

Procedure

Measure 50cm^3 of BA_1 and put it into a conical flask. Weigh 4g of substance P and put it in the conical flask containing BA_1, start the stop clock at once, collect the gas formed in a gas syringe fitted on the flask. Record the volume of the gas collected after 20 seconds.

Values obtained using BA_1

Time (s)			
Volume of gas (cm^3)			

Repeat the above experiment using 50cm^3 of BA_1 and 4g of substance P and record the volume of gas collected for the same time interval and tabulate the results differently.

Values obtained using BA_2

Time (s)			
Volume of gas (cm^3)			

Questions

(a) Plot a graph of Volume of gas collected against time for the two experiments.

(b) Where is rate of reaction high?

(c) Explain the appearance of the two graphs.

Note: The volume of gas per unit time gives the rate of reaction at any instant. Other factors such as temperature and pressure may affect the rate of reaction.

For example, an increase in temperature favors an endothermic reaction and decrease in temperature favors an exothermic reaction, this is according to Le' Chatelier's principle, e.g

$$N_2 \text{ }_{(g)} + 3H_{2(g)} \rightarrow 2NH_{3 \text{ }(g)} \quad \triangle \quad \text{-ve}$$

This reaction is exothermic and   therefore would not be favored by high temperature but high temperature would favor the decomposition of NH_3 into N_2 and H_2 i.e reverse reaction. Generally temperature increases the rate at which molecules move and collide per second.

Revision Questions:

1) Calculate the basicity of an Acid, HxM, IF $17.5cm^3$ of 0.1M of acid completely neutralizes $7.0cm^3$ of 0.5M KOH solution

2) 0.081g, of a metal oxide XO, was dissolved in $80cm^3$ of 0.05M H_2SO_4. The resultant solution that contained excess acid required $10cm^3$ of 0.4M NaOH solution for complete neutralization. Calculate the relative formula mass of XO and hence the atomic mass of X (O=16)

3) If $25cm^3$ of a metal hydroxide $X(OH)_2$, solution made by dissolving 2.6g of $X(OH)_2$ in 1 liter of water required $23cm^3$ of 0.1M Nitric acid for completer neutralization. Calculate the atomic mass of X (O=16, H= 1)

4) If $20cm^3$ of an organic acid X of Molecular formula $H-(CH_2)nCOOH$, made by disolving 17.6g of the acid in 1 liter of water required $20cm^3$ of 0.2M NaOH for complete neutralization. Calculate the molar mass of X and deduce the value of n and hence the formula of X.

5) $18.75cm^3$ of 0.2M sodium hydroxide neutralized $25cm^3$ of 0.05M solution of the acid. Calculate
   (a) The number of moles of Sodium hydroxide that reacted
   (b) The number of moles of the acid that reacted.
   (c) The molar ratio of alkali to acid for the reaction

6) In an experiment to standardize a solution of sulphuric acid that was about 0.1M, 2.7g of pure anhydrous sodium carbonate was dissolved to make $250cm^3$ of Solution. $20cm^3$ of this solution needed $23.5 cm^3$ of the acid when titrated Using Methyl orange indicator. Calculate the concentration of the acid in g/l. If

800cm^3 of the acid were left after titration, how many cm^3 of water must be

Added to this Solution to make it exactly 0.1 M

7). In an experiment to determine the stoichiometry of neutralization, the following solutions were provided;

Hydrochloric acid of concentration 10.95g/l, and a solution-containing 6g of basic substance Q in 1 liter of water. If 25cm^3 of the acid solution were neutralized by 20cm^3 of solution of substance Q, using phenolphthalein indicator,

    (a) Calculate the molarities of hydrochloric acid, and substance Q

    (b) Calculate the moles of hydrochloric acid that reacted

    (i) The moles of substance Q that reacted.

    (ii) The ratio in which the acid reacted with Q

(Given that H= 1, Cl = 35.5, and the relative formula mass of Q = 59.5)

NOTE: The weight of the solute in a given amount of water is known as "concentration". This concentration can be expressed by two methods

(i)   The weight of the solute dissolved in a given volume of solution

(ii)   The weight of solute dissolved in a given volume of water

The weight of solute dissolved in a given volume of solution can be expressed in 3 ways, which include, NORMALITY, FORMALITY and MOLARITY. However, never use normality and formality. Always use Molarity as looked at in this book.

The sum of the atomic weights of individual elements indicated in the formula gives the molecular weight of that substance and should always be expressed in grams.

This weight is very important when calculating Molarity of a given substance. A solution can be called a molar solution if its Molarity is 1.

All chemical equations obey the law of conservation of mass (There must be equal masses on each side of the chemical equation). Ensure that the equation is well balanced and arranged in terms of atoms, molecules, ions, and electrons. This will help you to get the correct reaction ratio for example,

$$2NaOH_{(aq)} + H_2SO_{4\ (aq)} \rightarrow Na_2SO_{4\ (aq)} + 2H_2O_{(l)}$$

The above equation shows that, 1 mole of Sulphuric Acid reacts with exactly 2 moles of Sodium hydroxide to give 1 mole of sodium sulphate and 2 moles of water. Or 80g of $NaOH$ + 98g of H_2SO_4 will give 142g of Na_2SO_4, + 36g of H_2O. If we can compare both sides of the reaction, you will see that, there are equal masses on each sides of the reaction. So the above equation obeys the law of conservation of mass.

PART FOUR

QUALITATIVE ANALYSIS

Qualitative analysis is a technique used to identify the different ions that could be contained in a given compound. One is required to observe the compound given to him /her to identify its color if its colored, its textures, and its general appearance. This will act as a starting point for further analysis.

The physical properties which include color, smell, appearance and solubility of a compound in water will greatly help a candidate to predict which ions could be contained in that compound even before other tests are carried out.

Observing the texture of substance given to you is very important in a way that you are able to deduce whether the substance is a liquid, solid, hydrated, crystalline, or powder. All these are very important observations during preliminary analysis of compounds.

If heating is required, always be very careful of the gases evolved. Some gases are poisonous and therefore should not be drawn directly or closer to the nose. Gases can be identified using damp litmus paper(s), a glowing splint, smell and their confirmatory tests. It's therefore important for a candidate to know all the confirmatory tests for different gases before carrying out a heating test.

Generally qualitative analysis of inorganic compounds is divided into 4 stages.
1. Preliminary test
2. Making a solution or heating the substance under test
3. Testing for cations
4. Testing for anions

The last two stages are sometimes referred to as confirmatory tests.

The purpose of preliminary tests is to provide some general pointers as to the nature of the substance under observation. Different ions show different characteristics in different reagents as discussed below.

Finally we carry out confirmatory tests to confirm the presence of those particular ions, which could have been suspected in the preliminary tests prior. However, more than one reagent can be used to confirm the presence of an ion. It is therefore important for a candidate to know how a particular ion reacts with these different reagents used. Regular practice of these test experiments will help a candidate to become familiar with all the reagents used and the different characteristics of the ions in these reagents.

4.1 Appearance of Substance

As already discussed in the introduction above, one can use the appearance of a substance to tell what it could comprise. The following rules can guide somebody to identify the components of any substance before other preliminary tests are carried out.

4.1.1 Color

The color of a substance may give a clue to its identity. The table below lists some cations and their colored compounds when dissolved in water.

Ion	Main color of solution	Examples of compounds
Cu^{2+}	Blue	$CuSO_4.5H_2O$
Fe^{3+}	Yellow/brown	$FeCl_3.6H_2O$
Fe^{2+}	Pale green	$FeSO_4.7H_2O$
Cr^{3+}	Green	
Ni^{2+}	Green	$NiSO_4.6H_2O$

Mn^{2+}	Pale pink	
$Cr_2O_7^{2-}$	Orange/red	$Kr_2Cr_2O_7$
CO_4^{2-}	Yellow	$Kr_2Cr_2O_4$

Summary

Color	Inference
White or colorless	Probably substance does not contain a transition metal
Substance colored	Probably it contains a transition metal
Substance colored blue	Possibly Cu^{2+}, or Ni^{2+}
Substance colored green	Possibly Ni^{2+}, Fe^{2+} Cu^{2+}, or Cr^{3+}
Substance colored yellow	Possibly Fe^{3+}

Compounds of a particular ion have same general color and they display the same color in aqueous solution as shown above.

4.1.2 Texture and Physical Appearance

Carbonates and oxides are usually powdery and do not appear as crystalline. Most inorganic substances are usually high melting point solids where as organic substances though not covered in this book are often liquids or solids with low melting points.

4.1.3 Smell.

Inorganic substances posses particular smells e.g Ammonium carbonate posses smell of ammonia. Bleaching powder posses smells of Chlorine.

Note:

1) Colored compounds are characteristic of transition metals. However, Ag+ and Zn^{2+} are colorless.

2) Color alone cannot be used to identify and confirm the presence of a given metal iron. One has to carry out the necessary preliminary and confirmatory tests so as to fully confirm the presence of a metal.

3) Metal cations that are non hydrated are often colorless.

There are some transitional metal ions that show variable oxidation states these ions can show different colors in their different oxidation states. For example, Fe^{2+} is green while F^{3+} is brown.

4.2 Effect of Heating on Solid Substances

Heating of solids should not be carried out in ordinary test tubes; solids should be heated in a dry hard glass tube.

When heating a substance, be careful at the very beginning, look out for water vapor condensing on the cold parts of the tube.

The most important one must look out for while heating, is the evolution of a gas, which can be tested with either litmus paper, or other appropriate reagents like limewater. Observe the appearance of the residue while heating and then after heating. i.e when residue is cold. Residues of different metal oxides show different characteristic colors upon heating and on cooling.

Table showing most common gases and their sources

Gas	Source
NH_3	Ammonium compounds
CO_2	CO_3^{2-}, HCO_3, and some oxalates
Cl_2	Hypochlorites, unstable chlorides
Water vapor	Damp hydrated salts
O_2	Peroxides, ClO_3
NO_2	NO_3, but not those of K, Na, and NH_4,
SO_2	Sulphates thiosulphates and sulphites
H_2S	Hydrogen sulphide

Note: always use a dry hard (ignition) glass test tube to heat a solid substance and use a small quantity. Do not put much of the substance given to you. Heat until there is no further change then use the color and smell to identify the gases evolved (if any). It is very important to always use damp litmus paper even when not instructed to do so. Identify the gases given out. Observe the appearance of the residue left after heating.

4.1.4   Appearance of Residue

Be very careful with this technique of analysis as it may be misleading because the residue may quickly change color upon cooling. However, the following inferences may be helpful if critical observations are made.

Appearance of Residue	Inference
Red brown residue produced from heating a green solid	Ni^{2+}, F^{2+} salt originally present
Black hot, reddish brown when cold	Fe_2O_3 present
Yellow when hot, white when cold	ZnO produced from breakdown of Zn salt
White residue produced after heating a blue solid	$CuSO_4$
Brown when hot, yellow when cold	PbO

4.3 Solubility of Salt in Cold water

When instructed to add cold water to a substance, always observe whether the substance completely dissolves or whether it partially dissolves in water. Solubility of a salt in water is very important. However, there are some substances, which do not dissolve in water (insoluble) as shown in the table below. Most students forget to shake properly once instructed to dissolve the solid substance in water and still deduce that the substance is insoluble. This is a very important stage of analysis as seen at the beginning of this part.

Once you fail to make a solution or mixture using water as could have been instructed, then you will not satisfactorily analyze the components of substance you could be dealing with. Do not heat if the solid fails to dissolve unless instructed to do so.

The table below will help you identify the ions that completely and partially dissolve in water and those that are insoluble in water.

Anion	Soluble	Relatively soluble	Insoluble
CO_3^{2-}	Na^+, K^+, NH_4^+		The rest
NO_3^-	ALL		
SO_4^{2-}	The rest	Ag^+, Ca^{2+}	Pb^{2+}, Ba^{2+}, Ca^{2+}
PO_4^{2-}	Na^+, K^+, NH_4		The rest
SO_3^{2-}	The rest		Pb^{2+}, Ba^{2+}, Ca^{2+}
Halides	The rest	Pb^{2+}	Ag^+, Pb^{2+}, Cu^{2+}
CrO_4^{2-}	The rest	Ca^{2+}	Ag^+, Pb^{2+}, Ba^{2+}

4.4 Reactions of Cations With Major Reagents

This part gives the reactions of the common reagents with each cation. However, there are some cations that do not show any reaction on treatment with reagents. Some reactions only occur under certain conditions. Majorly the reactions looked at here apply to aqueous solutions of the Cation

4.4.1 Reaction with Sodium Hydroxide Solution.

Sodium hydroxide solution precipitates the insoluble hydroxides but the amphoteric hydroxides will re-dissolve in excess sodium hydroxide

Cation	Few drops	Excess
NH_4^+	NH_3 gas evolved	NH_3 gas evolved basically on warming
Ba^{2+}	No ppt	White ppt
Ca^{2+}	White ppt if solution is concentrated	White ppt insoluble
Mg^{2+}	White ppt	White ppt insoluble
Al^{3+}	White ppt	White ppt soluble due to formation of complex
Zn^{2+}	White ppt	White ppt soluble due to formation of complex ion
Pb^{2+}	White ppt	White ppt
Fe^{2+}	Dirty green ppt slowly turns to brown	Green ppt soluble can be oxidized by air turning brown
Fe^{3+}	Rust brown ppt	Brown ppt soluble
Ni^{2+}	Green ppt	Green ppt insoluble
Cu^{2+}	Blue ppt	Blue ppt insoluble turning black if heated
Mn^{2+}	White ppt	Insoluble turns brown on standing
Cr^{3+}	Grey ppt soluble	

Summary of Color of ppts With 2M NaOH

Color of ppt	Suspected cation
Pale blue	Cu^{2+}
Green ppt turning brown	Fe^{2+}
Rust brown ppt	Fe^{3+}
Grey-green ppt	Cr^{3+}
Pale brown ppt	Ag^{+}
Green ppt	Ni^{2+}
Cream ppt turning dark brown	Mn^{2+}

4.4.2   Reaction with Ammonia Solution

Ammonia reagent precipitates insoluble metal hydroxides. Some hydroxides precipitates dissolve in excess Ammonia solution due to formation of complex ions (soluble amine complexes)

Cation	A few drops	Excess
NH_4^{+}		
Al^{3+}	White ppt	Slightly soluble
Pb^{2+}	White ppt	insoluble
Zn^{2+}	White ppt	White ppt soluble giving a colorless solution
Cu^{2+}	Blue ppt	Blue ppt soluble forming a deep blue solution
Fe^{2+}	Dirty green ppt	Dirty green ppt insoluble forming a brown ppt on

		oxidation by air
Fe^{3+}	Rust brown ppt	Rust brown ppt insoluble
Ni^{2+}	Green ppt	Green ppt soluble giving a blue solution
Cr^{3+}	Grey ppt, green ppt	Slightly soluble forming pink solution
Mn^{2+}	White ppt slowly turns brown	White ppt insoluble slowly turns brown
Ba^{2+}		
Ca^{2+}		
Mg^{2+}		
Co	Blue ppt	Blue ppt soluble turning pink

4.4.3   Reaction with Dilute Sulphuric acid (H₂SO₄)

Here, the main reaction occurring is the precipitation of insoluble metal sulphates. However, some anions also react with dilute Sulphuric acid.

Ion	Observation	Reactions Occurring
Pb^{2+}	White ppt	$PbSO_4$ (ppt)
Ba^{2+}	White ppt	$BaSO_4$ (ppt)
Ca^{2+}	White ppt	$CaSO_4$ (ppt)
Ag^+	White ppt	$AgSO_4$ (ppt)

Note: $AgSO_4$ and $CaSO_4$ will precipitate if only the ion solution is concentrated

4.4.4 Reaction with Dilute Hydrochloric Acid

The main reactions occurring here is the precipitation of the insoluble metal chlorides (some anions also react with dilute hydrochloric acid)

Ion	Observation	Reaction occurring
Pb^{2+}	White ppt, slightly soluble in concentrated HCl	$PbCl_2$ ppt. Dissolves in concentrated HCl due to formation of $PbCl_2$ complex ion.
Ag^+	White ppt, which turns blue black on exposure to light. ppt dissolves in dilute NH_3 to give colorless solution	AgCl ppt. ppt decomposes in light to Ag metal (blue black) and chlorine

Note: ON adding dilute acid to the solid, always identify any gases evolved. The formation of a clear solution indicates absence of an insoluble metal chloride or sulphate.

Table showing effect of Acids on Solids

Observation	Deduction
Carbon-dioxide gas given off	Carbonates or bicarbonates (CO_3^{2-} or HCO_3^-) probably present
Chlorine gas given off	Hypochlorate probably present
Sulphur-dioxide	sulphite

4.4.5   Reaction with Sodium Carbonate Solution

All cations form ppts with sodium carbonates solution except for group 1 and Ammonium. The precipitates may be hydroxides, carbonates or mixtures. Usually the precipitates dissolve in dilute Acids often with evolution of carbon-dioxide gas, sodium carbonate is a strong alkali and thus gives off ammonia from ammonium salts basically on warming.

Ion	Observation	Reaction occurring
Mg^{2+}	White ppt	$Mg^{2+} + CO_3^{2-} \rightarrow MgCO_3$ ppt
Ca^{2+}	White ppt	$Ca^{2+} + CO_3^{2-} \rightarrow CaCO_3$ ppt
Ba^{2+}	White ppt	$Ba^{2+} + CO_3^{2-} \rightarrow BaCO_3$ ppt
Al^{3+}	White ppt which dissolves in excess Na_2CO_3 solution	$Al(OH)_3$ ppt. the ppt dissolves in excess Na_2CO_3 due to formation of complex - $[Al(OH)_4]^{2-}$
Pb^{2+}	White ppt	$Pb^{2+} + CO_3^{2-} \rightarrow PbCO_3$ ppt
Mn^{2+}	Light yellow brown ppt	$Mn^{2+} + CO_3^{2-} \rightarrow MnCO_3$ ppt
Cr^{3+}	Blue-green ppt, dissolves in excess Na_2CO_3	
Fe^{2+}	Dirty green ppt slowly turns brown	$Fe(OH)_2$ is oxidized by air to $Fe(OH)_3$
Fe^{3+}	Rust brown ppt	$Fe(OH)_3$ ppt
Cu^{2+}	Pale green ppt	$Cu^{2+} + CO_3^{2-} \rightarrow CuCO_3$ ppt
Ni^{2+}	Green ppt	$Ni^{2+} + CO_3^{2-} \rightarrow NiCO_3$ ppt
Zn^{2+}	White ppt	$Zn^{2+} + CO_3^{2-} \rightarrow ZnCO_3$ ppt
Ag^+	White ppt	
NH_4^+	NH_3 given off on warming	$NH_4^+ + OH^- \rightarrow NH_3 + H_2O$

The reactions of the metal cations with Ammonium carbonate and sodium carbonate are similar. However, large amounts of Cu^{2+}, Zn^{2+} and Ni^{2+} ppts dissolve in Ammonium carbonate than in sodium carbonate.

4.4.6   Reaction with Hydrogen Peroxide (H_2O_2)

Hydrogen peroxide is a strong oxidizing agent in alkali solution. Metal cations with more than one oxidizing states are oxidized by H_2O_2 to their higher oxidation states. This reagent is normally added after NaOH solution.

Ion	Observation
Fe^{2+}	Green ppt, immediately turns to brown
Mn^{2+}	White ppt, immediately turns brown
Pb^{2+}	White ppt, turns brown if excess NaOH is added
Cr^{3+}	Green ppt, turns yellow forming a yellow solution

Note: ppt is an abbreviation of the word precipitate. It is used due to insufficient space provided in practical questions. It's not allowed to be used in answering theory questions, though it may be accepted in this particular part.

4.5 Confirmatory Tests for Cations

Metal cations have more than one confirmatory test. It is always very important for one to be familiar with all reagents used to confirm the presence of the different metal cations in a solution. Below are the common cations and the reagents used to confirm their presence.

Ion	Reagent	Observation and reaction occuring
Cu^{2+}	1. Add excess ammonia solution 2. Potassium hexacyanoferrate ii solution	1. A deep blue ppt of $Cu(OH)_2$, dissolves in excess. 2. Red –brown ppt
Ni^{2+}	1. Add few drops of Ammonia solution, then add dimethylgloximine solution 2. Potassium cyanide	1. Bright red ppt 2. A yellow green ppt of $Ni(CN)_2$ ppt dissolves in excess to form a dark yellow solution.
Fe^{2+}	Add potassium hexacyanoferrate III solution	Deep blue ppt of $Fe^{2+}(CN)_6$
Fe^{3+}	1. Add potassium hexacyanoferrate III solution 2. Add ammonia or thiocyanate solution	1. Deep blue ppt of $Fe^{3+}(CN)_6$ 2. Intense blood red color
Mn^{2+}	Add little dilute HNO_3 followed by a little sodium bismuthane	Purple color is produced in the solution, the purple color

		disappears after some time due to oxidation of Mn^{2+} to MnO_4
NH_4^+	Warm with NaOH	NH_3 gas evolved
Pb^{2+}	Add potassium chromate followed by dilute HCl	Yellow ppt of $PbCrO_4$ which reacts with HCl forming an orange solution
Ba^{2+}	Add potassium dichromate solution	Yellow ppt of $BaCrO_4.$ This dissolves in dilute Hcl giving a clear solution
Cr^{3+}	Add NaOH followed by H_2O_2 solution and warm gently. Acidify with dilute H_2SO_4	Yellow solution of Sodium Chromate formed. On addition of acid, a deep blue color is formed.
Mg^{2+}	Add a little solid NH_4Cl followed by a little ammonia solution shake to dissolve and add $NaHPO_4$ solution.	White ppt of Ammonium magnesium phosphate is produced.
Ca^{2+}	1. Potassium dichromate (K_2CrO_4) solution 2. Ammonium oxalate solution 3. Ammonium carbonate solution	1. No ppt (distinction from Ba^{2+}) 2. A white ppt of calcium oxalate, insoluble in hot dilute acetic acid 3. White ppt of calcium carbonate
Al^{3+}	1. Acidify the solution of suspected aluminium salt with dilute HCl, then add 4 drops of aluminium reagent followed by NaOH drop wise 2. Sodium carbonate solution	1. A pink lake is observed 2. White ppt of Aluminium hydroxide soluble in excess reagent.

4.6 Confirmatory Tests for Anions

Ion	Test (reagent)	Observation and reactions occurring
NO_3^-	Add H_2SO_4 and $FeSO_4$ solution shake then add conc. H_2SO_4 downside of the test tube.	A brown ring is observed at the surface of the mixture in the test tube.
SO_4^{2-}	Add dilute HCl followed by $Ba(NO_3)_2$ solution. Also HNO_3 may be added instead of HCl.	White ppt of $BaSO_4$
CO_3^{2-}	Add dilute HCl or HNO_3	CO_2 gas evolved this can be tested with lime water
NO_2^-	1. Add HCl 2. Add dilute H_2SO_4 and $FeSO_4$ solution	1. Effervescence and brown fumes of NO_2 in cold 2. Deep brown color in cold
$S_2O_3^{2-}$	Add dilute HCl to solution of suspected anion	SO_2 gas evolved and yellow white ppt of sulphur produced.
Cl-	Add $AgNO_3$ solution	White ppt of AgCl, ppt dissolves in dilute ammonia solution
Br-	Heat solid of suspected Br⁻ with conc. H_2SO_4.	HBr gas and brown Br_2 gas evolved. HBr gives dense white fumes with NH_3
I-	Heat solid with Conc. H_2SO_4	HI gas produced and black (I_2) solid observed when

		heated strongly . Violet fumes of I_2 gas observed
PO_3^{3-}	Acidify with conc Nitric acid and warm solution with quantity of ammonium molybodinate solution	White ppt
SO_3^{2-}	Warm solution of anion with dilute HCl	SO_2 gas evolved, no ppt of sulphur produced (distinction from $S_2O_3^{2-}$)
CrO_4^{2-}	1. Add AgNO$_3$ solution, 2. H$_2$SO$_4$ then add H$_2$O$_2$	1. Red ppt of Ag$_2$CrO$_4$ 2. Deep blue color which fades away very quickly.

4.7   EXPERIMENTS

Experiment 1.

You are provided with substance Q that contains one cation, and one anion. Carryout the following tests to identify the cation and the anion in Q. Record your observations and deductions in the table below.

Test	Observations	Deductions
To a spatula end full of Q in a boiling tube is added about $5cm^3$ of water and then the resultant solution divided into 4 portions	Q dissolves in water to give a light green solution	Q contains transition elements, probably Ni^{2+}, Fe^{2+}, Cu^{2+}, or Cr^{3+} present
To the 1st portion, add aqueous NaOH drop wise until in excess	Green ppt insoluble in excess NaOH	Probably Ni^{2+}, Fe^{2+} present
To the 2nd portion add dilute Ammonia solution drop wise until in excess	Green ppt which dissolves in excess ammonia solution to form a blue solution	Ni^{2+} confirmed
To the 3rd portion add silver nitrate solution drop wise until in excess	White ppt insoluble in excess reagent	Cl-, SO_4 probably present
To the 4th portion, add dilute Nitric acid followed by Barium nitrate solution	A white ppt insoluble in excess	SO_4^{2-} confirmed
Identify the cations and anions in Q. Cation: **Ni^{2+}** Anion **SO_4^{2+}**		

Experiment 2.

You are provided with substance Z, which contains three cations, and one anion. You are required to carry out tests so as to confirm the cations and the anions present in Z.

Test	Observations	Deductions
One spatula end full of Z in a test tube was heated.	A colorless gas was evolved, it turned moist red litmus paper red, and lime water milky.	Probably HCO_3^- and CO_3^{2-} present the gas is CO_2
To one spatula end full of Z was added $10cm^3$ of water and the mixture was shaken and the resultant solution was dissolved into 3 equal portions.	The substance dissolved completely leaving no residue, it formed a colorless solution	Z is probably a hydrogen carbonate of a non-transition element e.g K^+, Na^+, Li^+
To the 1st portion was added an aqueous solution of NaOH	There was no observable color change	Probably K^+, Na^+, Li^+ present
To the 2nd portion was added dilute HCl	Effervescence occurred. A colorless gas that turned moist blue litmus paper red and lime water milky was evolved	HCO_3^- and CO_3^{2-} suspected
To the 3rd portion was added a solution of aqueous magnesium chloride	There was no observable change	HCO_3^- confirmed
Identify the cations and anions in Z Cation: **Na^+, K^+, Li^+** Anion **HCO_3^-**		

Experiment 3

You are provided with substance M, which contains one cation and one anion. You are required to carry out the following tests to identify the ions present

Test	Observations	Deductions
A little sample of M was put in a clean spatula and burnt	M was a white powder that turned blue in moist air. It burnt with a bluish green flame	A hydrated salt suspected probably Cu^{2+} present
2spatula end full of M were dissolved in about $10cm^3$ of water and resultant solution divided into 4 portions.	Blue solution formed	Cu^{2+} Present. Probably salt of Cu^{2+}, e.g SO_4^{2-}, NO_3^{-} present
To the 1st portion was added 2 drops of NaOH, mixture filtered and residue was put in a dry test tube and heated until there was no further change.	Blue ppt was formed. Blue ppt was left on filter paper. Residue turned black on heating	Cu^{2+} Present $Cu^{2+}_{(aq)} + 2OH_{(aq)} \rightarrow Cu(OH)_2$ $Cu(OH)_2$ decomposes on heating to CuO. $Cu(OH)_{2(aq)} \rightarrow CuO_{(s)} + H_2O_{(l)}$
To the 2nd portion was added 3 drops of Na_2CO_3 solution and heated	Light blue ppt formed which darkened on heating and finally turned black. A colorless gas that turned moist blue litmus paper red and lime water milky	Cu^{2+} ions present $CuCO_3$, Precipitated decomposing on heating to CuO. CO_2 evolved
To the 3rd portion was	White ppt formed, ppt was	SO_4^{2-} confirmed.

added dilute HCl followed by barium Chloride solution	insoluble in excess HCl	$BaSO_4$ precipitated $Ba^{2+}_{(aq)}+ SO_4^{2-}_{(aq)} \rightarrow BaSO_{4\,(s)}$
To the 4th portion was added aqueous ammonia drop wise until excess	Pale blue ppt which dissolved in excess ammonia to form a deep blue solution	Cu^{2+} confirmed $Cu(OH)_2$ precipitated, dissolved in excess aqueous ammonia to form $Cu(NH_3)_2^{2+}$ Complex
Identify the cations and anions present in M Cation: Cu^{2+} Anion SO_4^{2-}		

Note:

(i) You can differentiate between SO_4^{2-} and HSO_4^- using Na_2CO_3 e.g

To the solution, add Na_2CO_3	Vigorous effervescence of CO_2 indicates HSO_4^-, slight or no effervescence indicates SO_4^{2-}

(iii)   You can differentiate between CO_3^{2-} and HCO_3^- using $MgSO_4$ solution. E.g

Add 2 drops of $MgSO_4$ solution to the solution of unknown.	A white ppt of $MgCO_3$ indicates CO_3^{2-}. No ppt, indicates HCO_3^-

Experiment 4.

You are provided with substance X, which contains one cation, and one anion. Carry out the following tests and identify the ions present.

Test	Observations	Deductions
To 2 spatula end full of X in a test tube was added conc. H_2SO_4 and warmed	X was a pale green crystal on addition to H_2SO_4, effervescence occurred. Gas evolved that formed white fumes with ammonia	Fe^{2+}, Cu^{2+} ion suspected. Cl^- ion present. HCl gas evolved. $HCl_{(g)} + NH_{3(g)} \rightarrow NH_4Cl_{(g)}$
To 2 spatula end full of X in a clean test tube was added MnO2 and mixed with conc. H2SO4 was added to the mixture and warmed.	Effervescence occurred, yellowish green gas evolved, gas bleached litmus paper	Cl^- ion present, gas evolved is Cl_2
2 spatula end full of X were dissolved in about 10cm3 of water and the resulting solution divided into 3 portions	Light green solution was formed	Fe^{2+} ions suspected
To the 1st portion was added NaOH solution dropwise until in excess	Dark green ppt formed. Ppt insoluble in excess and slowly turned brown	Fe^{2+} ions present $Fe^{2+} + 2OH^-_{(aq)} \rightarrow Fe(OH)_{2(aq}$
To the 2nd portion was added aqueous ammonia solution drop wise until in excess	Dirty green ppt formed. Insoluble in excess, ppt slowly turned brown	Fe^{2+} Present $Fe(OH)_2$ precipitated . Fe^{2+} oxidized to Fe^{3+} $Fe^{2+} + 2OH^-_{(aq)} \rightarrow Fe(OH)_{2(aq}$
To the 3rd portion was	White ppt formed.ppt	Cl^- ions confirmed. AgCl

added HNO₃ followed by AgNO₃ solution and aqueous ammonia solution	dissolved in aqueous Ammonia solution	precipitated
Identify the ions present in X Cation: **Fe^{2+}** Anion **Cl^{-}** The salt is therefore, FeCl$_2$		

Test Yourself

Assuming you are provided with substance Q that is a mixture of anhydrous copper sulphate and copper carbonate, state the observations and deductions that could be made if the following tests are carried out on Q. Also identify any gase(s) that may be evolved. Record your observations and deductions in the table below.

Test	Observation	Deductions
Heat a spatula end full of Q strongly in a dry test tube.		
Add 4-5 cm^3 of water to two spatula end full of Q, shake and filter the mixture, keep both the residue and the filtrate.		
Divide the filtrate into 4 portions of the filtrate. To the 1st portion, add dilute NaOH solution drop wise until in excess.		
To the 2nd portion of the filtrate, add dilute Ammonia solution drop wise until in excess		
To the 3rd portion of the filtrate, add 2-3 drops of lead nitrate solution		
To the 4th portion of the filtrate carry out a test of		

your choice.		
Wash the residue in a test tube and add dilute HCl drop wise to dissolve it. Divide the resultant solution into 2 portions.		
To the 1st portion of the solution, add dilute NaOH solution drop wise until in excess.		
To the 2nd portion add dilute Ammonia solution drop wise until in excess		

Identify the cation and anions in Q

Cations……………………………………….

Anions……………………………………….

APPENDICES:

APPENDIX 1: TABLE SHOWING ATOMIC SYMBOLS, NUMBERS, AND WEIGHTS OF COMMON ELEMENTS

Element	Symbol	Atomic number	Approximate Atomic Weight
Hydrogen	H	1	1
Helium	He	2	4
Lithium	Li	3	7
Beryllium	Be	4	9
Boron	B	5	11
Carbon	C	6	12
Nitrogen	N	7	14
Oxygen	O	8	16
Fluorine	F	9	19
Neon	Ne	10	20
Sodium	Na	11	23
Magnesium	Mg	12	24
Aluminium	Al	13	27
Silicon	Si	14	28
Phosphorous	P	15	31
Sulphur	S	16	32
Chlorine	Cl	17	35.5
Argon	Ar	18	39.9
Potassium	K	19	39
Calcium	Ca	20	40
Chromium	Cr	24	52
Manganese	Mn	25	55

Iron	Fe	26	56
Nickel	Ni	28	59
Copper	Cu	29	64
Zinc	Zn	30	65
Bromine	Br	35	80
Silver	Ag	47	108
Iodine	I	53	127

APENDIX 2: PROPERTIES OF COMMON ELEMENTS

Name	Color at room Temperature	Common Oxidation number
Aluminium	Silver metal	+3
Argon	Colorless gas	0
Barium	Silver metal	+2
Beryllium	Silver metal	+2
Boron	Black solid	+3
Bromine	Red-brown liquid	-1, +5
Calcium	Silver metal	+2
Carbon	Black solid	+2, +4
Chlorine	Green yellow gas	-1, +5, +7
Chromium	Silver metal	+2, +3, +6
Copper	Red metal	+1, +2
Fluorine	Yellow gas	-1
Helium	Colorless gas	0
Hydrogen	Colorless gas	-1, +1
Iodine	Black solid	-1, +5
Iron	Silver metal	+2, +3
Lead	Silver metal	+2, +4
Lithium	Silver metal	+1
Magnessium	Silver metal	+2
Manganese	Silver metal	+2, +4, +7
Mercury	Silver liquid	+1, +2
Neon	Colorless gas	0
Nickel	Silver metal	+2
Nitrogen	Colorless gas	-3, +3, +5
Oxygen	Colorless gas	-2

Phosphorous	Yellow solid	+3, +5
Potassium	Silver metal	+1
Silicon	Black solid	+4
Silver	Silver metal	+1
Sodium	Silver metal	+1
Sulphur	Yellow solid	-2, +4,+6
Zinc	Silver metal	+2

APENDIX 3: ELECTROCHEMICAL SERIES
E. volts

$Li^+_{(aq)} + e \rightarrow Li_{(S)}$	-3.04
$K^+_{(aq)} + e \rightarrow K_{(S)}$	-2.92
$Ba^{2+}_{(aq)} + 2e \rightarrow Ba_{(S)}$	-2.90
$Ca^{2+}_{(aq)} + 2e \rightarrow Ca_{(S)}$	-2.87
$Na^+_{(aq)} + e \rightarrow Na_{(S)}$	-2.71
$Mg^{2+}_{(aq)} + 2e \rightarrow Mg_{(S)}$	-2.37
$Al^{3+}_{(aq)} + 3e \rightarrow Al_{(S)}$	-1.66
$Zn^{2+}_{(aq)} + 2e \rightarrow Zn_{(S)}$	-0.76
$Fe^{2+}_{(aq)} + 2e \rightarrow Fe_{(S)}$	-0.44
$Pb^{2+}_{(aq)} + 2e \rightarrow Pb_{(S)}$	-1.33
$H^+_{(aq)} + e \rightarrow \frac{1}{2} H_{2(g)}$	-0.00
$Cu^{2+}_{(aq)} + 2e \rightarrow Cu_{(S)}$	+0.34
$\frac{1}{2} I_{2(g)} + e \rightarrow I^-_{(aq)}$	+0.54
$Ag^+_{(aq)} + e \rightarrow Ag_{(S)}$	+0.80
$\frac{1}{2} Br_{2(g)} + e \rightarrow Br^-_{(ag)}$	+1.07
$\frac{1}{2} Cl_{2(g)} + e \rightarrow Cl^-_{(aq)}$	+1.36
$Au^+_{(aq)} + e \rightarrow Au_{(S)}$	+1.68
$\frac{1}{2} F_{(g)} + 2e \rightarrow F^-_{(aq)}$	+2.85

APPENDIX 4:COMMON IONS AND THEIR CHARGES

NAME	SYMBOL	CHARGE
Aluminum	Al^{3+}	+3
Ammonium	NH_4^+	+1
Barium	Ba^{2+}	+2
Calcium	Ca^{2+}	+2
Chromium III	Cr^{3+}	+3
Copper II	Cu^{2+}	+2
Iron II	Fe^{2+}	+2
Iron III	Fe^{3+}	+3
Lead II	Pb^{2+}	+2
Magnesium	Mg^{2+}	+2
Nickel	Ni^{2+}	+2
Potassium	K^+	+1
Silver	Ag^+	+1
Sodium	Na^+	+1
Zinc	Zn^{2+}	+2
Bromide	Br^-	-1
Chloride	Cl^-	-1
Cyanide	CN^-	-1
Fluoride	F^-	-1
Iodide	I^-	-1
Oxide	O^{2-}	-2
Hydroxide	OH^-	-1
Sulphide	S^{2-}	-2

Carbonate	$CO_3{}^{2-}$	-2
Chlorate	$ClO_3{}^-$	-1
Chromate	$CrO_4{}^{2-}$	-2
Dichromate	$Cr_2O_7{}^{2-}$	-2
Hydrogen carbonate	$HCO_3{}^-$	-1
Hydrogen sulphate	$HSO_4{}^-$	-1
Nitrate	$NO_3{}^-$	-1
Sulphate	$SO_4{}^{2-}$	-2
Phosphate	$PO_4{}^{3-}$	-3
Permanganate	$MnO_4{}^-$	-1
Chlorite	$ClO_2{}^-$	-1
Hypochlorite	ClO^-	-1
Nitrite	$NO_2{}^-$	-1
Sulphite	$SO_3{}^{2-}$	-2

Concluding Message From the Author

I have taught Chemistry subject in schools for years and I've seen how it is very easy to pass it, if a student only gets interest in the subject, while respecting the teacher. I developed interest in science and particular Chemistry subject before I reached secondary school. I liked seeing things happen, I liked adding salt or sugar to water, and watch it disappear in water. I could carefully watch gas bubbles coming out of boiling water and bursting on the surface of the water, it was fun to me seeing water change from liquid to gas, and then to drops of water on the cooler sides of the material covering the hot water. I did several experiments in my mother's kitchen even before I came to know of Chemistry subject. I added different amounts of salt to 1 cup of water, and watched how the different quantities of salt took different times to disappear in equal volumes of water. I could add paraffin to water and watch how the two liquids failed to mix, formed layers and yet paraffin-to-paraffin, or water-to-water mixed completely. It was all fun to me.

When I joined Secondary school, Chemistry subject looked a game to me. A number of experiments especially in grade one were not different from my daily games at home. I saw another game inform of a subject. Balancing chemical equations became fun to me, actually I got addicted to balancing chemical equations that I could want to generate an equation form anything I saw happening. Stubbornly I wanted to generate equations from events because I knew before an event; there was a force that contributed to that particular event. It was all fun to me. The only difference from this game (Chemistry subject) and my games in my mother's kitchen was that, I was awarded marks with a promise that, life could change through application of simple scientific concepts.

I could completely fail to understand why fellow students failed to pass this subject. I couldn't imagine any student failing to understand that paraffin and water are immiscible, or water appears in three different stages.

Many years have passed and I can see the value of this subject in my life. I would have said that, "The best way to predict your future is to create it" but looking back in the times, I can't say that I was surely creating my future through Chemistry. No, there are many ways through which people pass to get to a single point called destiny. You can find your route different from mine, and we still meet somewhere in life doing similar things.

Most important, if you have already identified yourself and chosen what you want in life, then you can be that what you want if you do it fully with passion and commitment. All the wishes if not based on valid principles will not produce the quality of life you want. It is not good to dream to become a doctor if you cant understand what it takes to become a doctor, it is not enough to just talk about it. All your efforts have to be based on practical realities that produce result and one of them is you understanding for example what it takes to see your dreams come to pass.

Chemistry is like any other subject that you love very much. You can only chose to dislike it, just because your thoughts have dictated upon you, and directed you to hate it. But those who have done it and liked it have determined that nothing will stop them from getting forward towards their life dreams. I accept that there are some topics, which you may fail to understand. But constant practice, consultations, and more research will make you leap over these hurdles and you will realize that these hurdles are just strengthening points, and yes you are on the way to your destiny.

It is also important for you to listen to people who have been where you want to go. If your interest is in becoming a scientist, a doctor, or a pharmacist, then know what it takes to reach there. Find out from someone who studied Chemistry. I had friends who were doing Chemistry subject at "A" Level, and I loved seeing them balancing Chemical equations. I interacted with them several times, and made them my friends. A few years later I found myself balancing chemical equations and teaching this subject.

Develop in depth learning. Do not cram for exams and forget everything two weeks after the exams. In-depth learners make the acquired knowledge part of their lives. They understand more about the world, their friends, and themselves. They use the acquired knowledge to build their world, and they make a difference in this world.

Never think of failure, remain focused and concentrate on your books. Your job as a student now is to read your books and understand everything you are being taught. That is your full time job.   Remember that you have a lot to cover but in a very short time. It is therefore up to you to properly divide your time so that you can balance all the subjects well.

Be highly encouraged and motivated all the time. Do not allow anything to discourage you. Put aside all negative thoughts and ensure that you only focus on your dreams. There are such things such as pornography and watching dirty movies instead of learning that can spoil your minds and take the time you would utilize to read your books. Avoid wasting your precious time on dirty things that will not yield you much. There are such power giving books as the Bible, which you could probably read through in your free time instead of spoiling your mind with dirty movies. You can only acquire "spiritual intelligence" by feeding your mind with good thoughts, and wisdom giving materials. And yes you can dream big again.

I am able to communicate to the world through my work because I am highly inspired, and my inspiration is from God, and this is why I am telling you NEVER to put God out of your life even when you are still a student. Love God, and honor Him as the giver of everything. Even if a subject is hard for you, lean on Him and ask Him to give you wisdom to understand only what you are meant to do because not all of us are called to

136

be scientists. Not all of us are called to be politicians. Ask God to give you wisdom to master only what you are supposed to master.

Remember your life is your best gift and your biggest investment; it is your best subject, and your precious capital. Guard it well. Preserve your life by all means. Avoid ALL things that could lead you to destruction. Do not go to dark places in dark hours; do not try to eat what you are not meant to eat. Do not try to drink what you are not mean to drink. Avoid eating anything, which you are instructed to take with a lot of care. Avoid it. Why risk your life?. Love to leave a legacy through the skills you have acquired in class. Use your wisdom to make this world a better place to live in. And remember that, our lives are a result of our choices. You can look at your life as a splendid torch, which you have to hold fast, and make it shine brightly as you prepare to hand it over to the future generations.

Hard work, faith, focus and commitment will overcome all seemingly impossible odds in your life.

Thank you for purchasing this product. Part of your money will go towards building our Church. May the good LORD Bless you in your endeavors.

Adams Guscan

www.ingramcontent.com/pod-product-compliance
Lightning Source LLC
Chambersburg PA
CBHW080659190526
45169CB00006B/2181